Soil and Rock Description
in Engineering Practice

Soil and Rock Description in Engineering Practice

David Norbury

Consultant; Director, David Norbury Limited, Reading, UK
Honorary Professor in Engineering Geology, University of Sussex, UK

WHITTLES PUBLISHING

CRC Press
Taylor & Francis Group

Published by
Whittles Publishing,
Dunbeath,
Caithness KW6 6EY,
Scotland, UK

www.whittlespublishing.com

Distributed in North America by
CRC Press LLC,
Taylor and Francis Group,
6000 Broken Sound Parkway NW, Suite 300,
Boca Raton, FL 33487, USA

Reprinted 2014

ISBN 978-1904445-65-4
US ISBN 978-1-4398-3634-7

Printed and bound by CPI Group (UK) Ltd, Croydon, CR0 4YY

To Susan
The rock without parallel

Contents

Preface

Geoscience graduates are routinely sent out on site with little or no training to carry out logging of soils and rocks as part of a site investigation. It is assumed that they have been taught how to carry out a systematic description in their university courses, but this is generally not the case. For this reason we established a course in Soil and Rock Description (SARD) at Soil Mechanics Limited in the mid-1980s. It arose from the critical review that I and colleagues carried out of Section 8 of BS 5930:1981 which led to our published recommendations (Norbury, Child and Spink, 1986). These courses which have now been delivered within Soil Mechanics over 100 times to date varied in length from half a day to four days, but it generally takes about three days to cover the material satisfactorily when practical sessions are included. This compares with the few minutes briefing and perhaps a few more minutes at the core box that most young graduates are given.

This book represents a distillation and refinement of the material assembled for the SARD course over the last 20 years. During this period, the standards framework has changed significantly. BS 5930 was revised in the 1990s to accommodate the experience gained in implementing the first version. At the same time, an ISO committee was drafting international practice which has since been published as normative EN ISO Standards around the world. These Standards introduced changes to descriptive terms which have had to be accommodated into practice. This led to the amendment of Section 6 of BS 5930 in 2007 and 2010.

The aim of this book is to provide practical guidance for those having to go out in the field to undertake engineering geological logging of the soil and rock samples and the exposures that are available. The systematic and codified approach is different from any other geological logging that most will have been taught and so care is required to ensure that the right terms are used in accordance with the definitions. A standard word order is provided not, as many believe, merely for the sake of it, but because if the descriptors are used in a consistent format life is much easier for all, mistakes are less likely and better communication is achieved.

The guidance within this book is not only for the young practitioners learning their craft however. It is also aimed at their seniors and mentors, including the responsible expert who is signing off the logs and reports on behalf of the company. These individuals have often been logging for many years, although in practice they may not have got their hands dirty for a while. They also need to catch up on the precise details of the current approach and the changes that have been introduced if they are to avoid mistakes which could be costly.

SARD is an art form in that the aim is to convey to the reader the key features of the ground that were visible in the exposure as though they were also present. This is important because the soil samples are not usually inspected by any party other than the driller and logger and in any case these deteriorate quickly and are thrown away within a few weeks. The field logs therefore provide the only record of the ground that is available to the designer and contractor later in the construction process.

I would like to note my thanks and debt to a number of individuals with whom I have worked in establishing and evolving the descriptive framework. In the early days, I acknowledge the input of Geoff Child and Tim Spink as we reviewed the 1981 Standard and prepared our paper for the Guildford conference in 1984. The in-house course was then developed with Tim and Dick Gosling. In particular, it was Tim who sketched out a number of diagrams and figures, which have evolved into figures within this text. Ever since the first course in 1988, these two colleagues and I have collaborated closely in preparing and delivering courses and in the preparation of national standards. Without their collaboration over two decades or more, the subject of soil and rock description would not have reached its current refined state. This collaboration and evolution will continue over the coming years as there is still work to be done to remove inconsistencies which remain.

There are also the helpful comments and assistance provided by other SARD teachers and delegates within Soil Mechanics and beyond who are too numerous to mention individually. The people who have attended the courses have helped over the years by challenging notions. This has required the preparation of substantiated responses. Finally, and far from least, thanks are due to Matt Wolsey for preparing all the line drawings in a form suitable for publication.

David Norbury

Definitions

The EN ISO standards start by providing definitions of many of the terms used in the standards. Provision of clarity at an early stage of the document is very useful. This lead is followed here and definitions of some key terms relevant to the description of soils and rocks are given below.

Description of soils and rocks is the reporting of the material and mass characteristics seen in the exposure or sample. The description is carried out in the field and does not require test results to be completed. Descriptions are presented as a major component of the field log and should be an interpretation which represents the in situ condition of the ground at the investigation position before any disturbance caused by the formation of the exposure or the taking of the sample.

Classification is a process that follows description and uses the geological succession, the variability of the engineering characteristics of the succession and the results of tests to place the soil or rock into pre-defined classes which are of uniform character and this assists in the evolution of the ground model. Classifications are not normally presented on field logs, but are used to present and discuss the ground conditions within the report and on plans and sections of the site.

Soils are assemblages of mineral particles and/or organic matter in the form of a deposit which can be separated by gentle mechanical means and which include variable amounts of water and air (and sometimes other gases). They include naturally deposited sediments and organic materials. Soils can be considered in terms of soil material and soil mass and display relict structure or fabric features when weathered in situ. The difference in properties between the material and the mass is not as large as in rocks, although the discontinuities in soil can also be critically important to the overall behaviour.

Rocks are naturally occurring assemblages of minerals which are consolidated, cemented or bonded together so as to form material generally of greater strength than soils. Rocks are composed of the **rock material** which is present between the discontinuities and the **rock mass** which is the whole rock together with the discontinuities and weathering profile. The discontinuities dominate the behaviour of the rock in engineering terms.

Rockhead is the boundary between materials that are described or classified as soil and those that are considered as rock. There are a number of definitions of rockhead, which can be based on strength, stiffness, structure or fabric, geological origin or specification clauses applied to the site works. In field description the identification of both the geological and engineering rockhead is usually necessary.

Made ground is the collective name for those soils which are laid down by man and are composed of either natural or man-made materials.

Marl is a calcium carbonate or lime-rich mud or mudstone which contains variable amounts of clays and aragonite. Marl was originally a term loosely applied to a variety of materials, most of which occur as loose, earthy deposits consisting chiefly of a mixture of clay and calcium carbonate, formed under freshwater conditions. The term is today often used to describe indurated marine deposits and lacustrine sediments which are probably more accurately named marlstones. Marlstone is an indurated rock of about the same composition as marl, more correctly called a clayey argillaceous limestone. It tends to have a blocky sub-conchoidal fracture and is less fissile than shale.

Loam is a soil composed of sand, silt and clay in proportions typically of about 40-40-20%, respectively, and is generally considered ideal for gardening and agricultural uses. Loams are gritty, moist and retain water easily. In addition to the term loam, different names are given to soils with different proportions of secondary constituents including sand, silt and clay (see Chapter 13.1.1).

The **nature of the soil grains** refers to their particle size grading, shape and texture or, if appropriate, their plasticity, together with special features such as organic or carbonate content. The nature of a soil does not usually change during civil engineering works. Nature can be described on disturbed samples.

The **state of the soil grains** is their packing, water content, strength or relative density and stiffness. The state of a soil usually changes during civil engineering work; the description of the soil state requires undisturbed samples or exposures.

The **fabric** or **structure** of a soil or rock comprises the large scale spatial interrelationship of elements and can include folds, faults, or the pattern of discontinuities defining the soil or rock blocks. The structure includes features that are removed by reconstitution, that is fabric or microfabric features, such as layering, fissuring or cementing. Structure is often destroyed by large distortions and so can be observed only in the field on natural or artificial exposures or, to some extent, in an undisturbed (Quality Class 1) sample.

Texture is the size, shape and arrangement of the grains (sedimentary rocks) and the crystals (igneous rocks).

Particle size grading is the proportion of particles of each size fraction by weight. Size fractions are defined on the basis of a range of particle size. Examples include clay or sand. The range of soil particles is divided into a number of size fractions. These range from boulders which are larger than 200 mm, through cobbles, gravel, sand and silt to clay. The individual particles in silt and clay fractions are not visible to the naked eye. Some size fractions are subdivided into coarse, medium and fine.

The **matrix** is the finer grained component of the soil or rock which fills or partially fills the voids between the grains which can also be called clasts or crystals. Soils or rocks with a range of size fractions can be either clast or matrix supported, indicating which size fraction is present as a continuum and which will therefore control the behaviour.

Discontinuities are the mechanical breaks which intersect the soil or rock material and break it up into fissure or joint bounded blocks. Examples include fissures (a term used in soils), joints, shears, faults and cleavage. The term discontinuity is synonymous with the term **fracture.**

Bedding planes separate different beds and represent changes in the depositional sequence. Bedding can be indicated by colour or material changes, but the latter are usually more important in engineering logging. Bedding planes are not necessarily a plane of weakness and therefore may not be a fracture.

Bedding fractures occur along bedding planes. Identifying whether a bedding fracture is induced by the sampling process or is naturally present in the ground can be very difficult.

Joints are breaks of geological origin in the continuity of a body of rock along which there has been no visible displacement. A group of parallel joints is called a set and joint sets intersect to form a joint system. Joints can be open, filled or healed. Joints frequently form parallel to bedding planes, foliation and cleavage and may be termed bedding joints, foliation joints and cleavage joints accordingly.

The spacing of joints tends to decrease with stress relief and weathering, that is as the depth of burial decreases. Fissures are exactly the same as joints but the term is usually reserved for such features in soil. This distinction between joints in rock and fissures in soil is not standardised, but has grown as a common practice.

Cleavage fractures occur along the preferentially aligned mineral that makes up cleavage. Care needs to be exercised in recording the spacing of cleavage fractures. Cleavage is a penetrative foliation and so the number of fractures present will tend to change with time. A freshly recovered core could well be recovered as a complete intact cylinder without any fractures being observable. As the core responds to the removal of the surrounding stress, dries out and is mechanically disturbed, fractures start to appear, such that within a few days the core will split into discs and these will become progressively thinner as more and more fractures appear.

Incipient fractures are natural fractures which retain some tensile strength and so may not be readily apparent on visual inspection.

Induced fractures are those created by the drilling, sampling or excavation process, which were not fractures in the ground. The assessment of these fractures is often not straightforward. The standard logging convention is that induced fractures are not described and are not included in assessment of the fracture state.

Faults are fractures or fracture zones along which there has been recognisable displacement, from a few centimetres to a few kilometres in scale. The walls are often striated and polished resulting from the shear displacement. The rock on both sides of a fault can be shattered and altered or weathered, resulting in fillings such as breccia and gouge. Fault widths may vary from millimetres to hundreds of metres. Faults can comprise single shear surfaces or multiple surfaces which combine to create a fault zone. The intervals between individual surfaces in a **fault zone** can be filled with **fault gouge** which is comminuted host material.

Shear surfaces are joints across which displacement has occurred. Shear surfaces tend to be smoother than joints and may have polished surfaces. Depending on a combination of confining pressure, grain size and joint spacing of the rock, and the amount of movement, shear surfaces can be striated.

The **discontinuity features** that should be routinely described in rock mass assessment are outlined in Chapter 11 and include:

- Orientation – attitude of the discontinuity described by the dip direction (azimuth) and dip of the line of steepest declination in the plane of the discontinuity.
- Spacing – perpendicular distance between adjacent discontinuities. Normally refers to the mean or modal spacing of a set of joints.
- Persistence – length of a discontinuity observed in exposure. May give a crude measure of the areal extent or penetration length of a discontinuity. Termination in solid rock or against other discontinuities reduces the persistence.
- Roughness – inherent surface roughness and waviness relative to the mean plane of a discontinuity. Both roughness and waviness contribute to the shear strength. Large scale waviness may also alter the dip locally.
- Wall strength – strength of the walls of a discontinuity. May be lower than rock block strength owing to weathering or alteration of the walls. It is an important component of shear strength if rock walls are in contact.
- Aperture – perpendicular distance between adjacent rock walls of a discontinuity, in which the intervening space is air or water filled.
- Infilling or filling – material that separates the adjacent rock walls of a discontinuity and that is usually weaker than the parent rock. Typical filling materials are sand, silt, clay, breccia, gouge, mylonite. Also includes thin mineral coatings and healed discontinuities, e.g. quartz, and calcite veins.
- Seepage – water flow and free moisture visible in individual discontinuities or in the rock mass as a whole.
- Number of sets – the number of joint sets comprising the intersecting joint system. The rock mass may be further divided by individual discontinuities.
- Block size – rock block dimensions resulting from the mutual orientation of intersecting joint sets and resulting from the spacing of the individual sets. Individual discontinuities may further influence the block size and shape.

1 Introduction

A site investigation lives or dies by its field logs
—Dick Gosling, personal communication

The description of soils and rocks in engineering practice forms a major and hugely significant input to one of the fundamental components of an investigation of the ground, namely the field log. The field log should record the materials and strata seen in a borehole, trial pit or any other type of investigation hole, or record an exposure such as a cliff or quarry face. The record presented as the field log (see Figure 1.1) includes descriptions of the strata encountered and a range of other field observations which provide additional information to that presented within the description alone. The descriptions within the field log are a basic element of the factual information that underpins the entire understanding and interpretation of the ground conditions at the site. The field log is therefore a basic building block in the compilation of the ground model. The field log is also all that remains after the investigation has taken place, the trial pit has been backfilled and the borehole samples or cores are no longer available. The field log therefore has a life well beyond that of the investigation report.

The purpose of this book is to provide the background and framework for the recording of exposures (which can include samples, cores, trial excavations or exposures) and for the communication of this record to the reader or user of the information. The users of field logs include clients, designers and contractors.

The descriptions contained within a field log are used in a number of ways and at various stages of the investigation:

- They provide the designers with information to guide the scheduling of field and laboratory tests.
- They provide a framework used in the interpretation of those test results and in the lateral and vertical interpolation or extrapolation of those results to the whole site.
- In many cases, the description is relied upon, because test results are either not available or cannot be used since representative samples may not be available or recoverable.
- The descriptions are a key element in the preparation of the permanent works design in terms of providing information that helps in understanding how the ground will behave and interact with the structure.
- The field logs and descriptions also provide information that is critical to the contractors who have to carry out the works. The aspects that a contractor will be interested in may centre on the site preparation or the temporary works so their interest will include aspects of how the ground will respond to

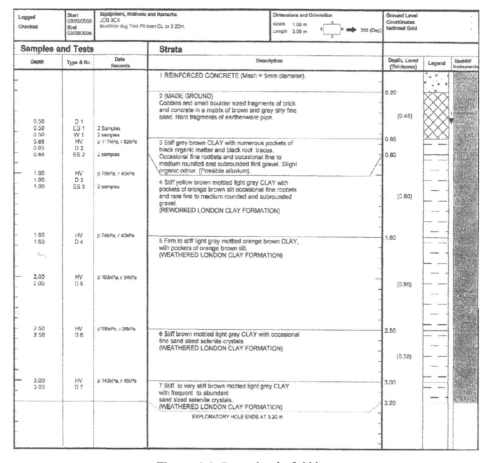

Figure 1.1 Example of a field log.

excavation, whether the materials are suitable for re-use, whether any ground treatment is needed and if so what sort might be appropriate and what support needs to be provided in the construction works.

For all these reasons it can be seen that complete and accurate description of the soils and rocks underlying a site is fundamental to all investigations. Compilation of the descriptions that appear on a field log is carried out by a field logging geologist or engineer. The completeness and correctness of the investigation report can therefore only be as good as this logger.

Despite the great importance which should thus be attached to field descriptions, this activity is usually carried out by the youngest members of staff, the newly qualified graduates. This is usually blamed on the need for a low price investigation. Whether this economy is false or genuine can be debated, but the true question is whether we as the investigation industry are providing the clients with a complete and professional service. All too often the answer to this question tends to be in the negative. Whilst many clients do not help themselves in this matter with their attitude to the cost benefits that can be achieved from an appropriate and professional site investigation, the industry should educate clients and manage the client's expectations.

Graduate staff who are deployed under this illusion of apparent economy may well have not received significant engineering geological training. The investigation industry tends to presume that these youngsters are trained and capable and expect them to launch immediately into description in the field. This presumption is usually made without reference to course curricula.

Even if these staff are geology graduates, they will probably not have received training specific to this most important activity, that is the engineering geological description of soils and rocks. The requirement of site investigation is for the systematic and accurate recording of all the samples or exposures using defined terminology in a standardised manner. Few undergraduate courses teach systematic description in a way that is sufficient for the investigation industry. There are exceptions, of course, but even these more enlightened establishments rarely teach the description of both soils and rocks.

Given these conditions, it is necessary for the requisite training to be provided after graduation and within industry. Until recently there have been no courses teaching soil and rock description on the open market and few companies provide formal in-house training courses. A large proportion of industrial training in this area is necessarily provided by one-to-one supervision or mentoring by more experienced members of staff within companies. Although this is a reasonable approach, it is far from certain that these staff have themselves received any formal training. The possibility that the young graduates will, therefore, be taught bad or even incorrect practices from the outset is significant and does happen. This is not helped by those individuals who believe that the best way to learn is by making mistakes. If this philosophy is invoked in soil and rock description (SARD) activities, the mistakes are quite likely to find their way into the field log on which the design work and the contractor's price are based.

The intention of this book is to provide a practical guide to carrying out soil and rock description in an engineering context for these juniors who do not have the necessary training or background experience. It also aims to fill gaps in the general level of knowledge about soil and rock description for the juniors and their seniors who check their work and sign off the reports and field logs that they include. In addition to the now current systematic practice, changes to practice that have taken place over the last 40 years and those that are taking place at the time of writing are highlighted. These changes are not all advances in descriptive practice, but they are now in normative standards and have to be followed until such time as those standards are revised and improved.

Throughout this book the process of the recording of exposures through description is given in detail. These exposures can vary from in situ material, such as a cliff face or trial pit or other in situ exposure, to continuous or intermittent individual samples recovered from a borehole and which are described remote from the field location. These samples can include disturbed and supposedly undisturbed samples and rotary drilled cores. The description of a number of such samples, be they large or small, leads to the compilation of a field log of the ground conditions. One purpose of this book is to provide guidance on the process of transferring the visual observations at the point of logging to the field log. This field log will be used by others and will remain as a record of the ground long after the samples have been discarded or the exposures covered up.

The reader should be aware that throughout all logging activities, there is a difference between the level of detail and the level of quality. A simple log is not detailed but can be of high quality because all the information that really matters is presented accurately. All field logs should be of high quality, but the level of detail to be recorded will vary depending on the nature of the ground and the requirements of the investigation. It is not appropriate to record the maximum level of detail in all investigations. The selection of the right level of detail requires experience together with a thorough understanding of the proposed works and how they and the ground will interact.

This book begins with a summary in Chapter 2 of the historical background to codification in soil and rock description over the last 50 years. This history is important as the terms and definitions in use have not remained unchanged and some terms have been recycled. Anyone using old field logs needs to be aware of these changes. In Chapter 3 the procedure for carrying out descriptions (which has not changed) is given, together with the standard word order which forms the basis for the systematic approach to this activity.

Description of soils and rocks is split into the description of the material features, which is the natural ground without discontinuities, and the mass which includes the discontinuities and the effects of weathering. Details of the description of the material are given in Chapters 4 to 9 in the same sequence as the standard word order. The description of the mass aspects of the ground is covered in Chapters 10 to 13.

The distinctions that have to be made to identify different classes of soil and rock, between coarse and fine soils and between inorganic and organic soils and between natural and anthropogenic soils is the first important step within a systematic description. The procedures for making these distinctions are outlined in Chapter 4, which provides guidance on the naming of the soils and rocks that are being described. Having named the material, description is then carried out in a sequence in accordance with the standard word order previously given in Chapter 3. The next chapters provide guidance on each aspect of the description in turn. Descriptions of the terms relative density, consistency and strength are given in Chapter 5 where the differences for each class of material require different approaches. Chapter 6 covers the description of the structure of the soil or rock, either through the description of fabric in soils, or of structure, fabric and texture in rocks. The whole of Chapter 7 is devoted to colour although for many this may overstate its importance. However, the need for accurate description of the colour is just as great as any other aspect of the material being described.

The behaviour of all soils and rocks is modified by secondary and tertiary constituents of the material, which need to be assessed in the field and reported. This is not always easy and although descriptions can be verified against subsequent laboratory test results, it is, of course, important that the constituent fractions are correctly assessed and described in the field. Guidance on this aspect of the description is provided in Chapter 8. The description of the soil or rock material is concluded with the identification of the geological formation and guidance on this aspect is given in Chapter 9. This is a relatively straightforward step, but care still needs to be exercised to get it right and not to overstate the confidence in the name assigned.

The mass features of all soils and rocks should be described because these aspects have a major influence on the behaviour of the ground in response to the engineering works. The description and classification of weathering is considered in Chapter 10. The logging of discontinuities in the mass is given in Chapter 11 with definitions of the terms to be used and the specific word orders to be adopted for this aspect of the description. Recording the fracture indices in the rotary core completes the recording of the discontinuities and this is covered in Chapter 12.

There are a number of other soil and rock types which also need to be described. Guidance on the description of low density organic and inorganic soils is given in Chapter 13 and on the description of made ground in Chapter 14.

All the guidance in the book up to this point relates to the description of the materials encountered. Chapter 15 discusses schemes for the classification of the soils and rocks which can be carried out at a number of different levels. Classification is the step that follows description and leads into the construction and evolution of the geotechnical model.

Finally the step-by-step process of actually carrying out a description is outlined in Chapters 16 and 17. The former chapter covers the description of boreholes by means of either intermittent samples or by continuous cores. The latter chapter covers the description of exposures in the field, whether these are natural or created as part of the investigation activity.

In terms of teaching the requirements of soil and rock description, this book can never replace a formal course where lectures are supplemented by practical sessions where delegates can get their hands dirty and really learn the procedures to be followed. However, the experience of delivering training courses in soil and rock description for over 20 years has enabled refinement of the content and coverage presented here and therefore enhances the usefulness of this book. The intention is to provide a background to systematic field description for practitioners both young and old, reminding readers of the terms that should be used and encouraging a higher standard across industry as a whole.

There have been a number of standards relating to geological description in engineering practice which includes BS 5930 in three versions, suggested guidance procedures from the international learned societies (International Association for Engineering Geology and the Environment, IAEG, and International Society for Rock Mechanics, ISRM) and also now international standards. The engineering geological profession has managed to progress from the position of no codification in the 1960s, to national, European and international codification which is the current position at the beginning of the 21st century. That this progression has been achieved within less than 40 years bears testament to the hard work of many practitioners from around the world in committing their time and energy to evolution in this area. The results of this work by many are captured in this book and due acknowledgement is paid to them collectively here.

1.1 What are we describing and why?

A key question to consider before starting any exercise involving description of elements of the ground is 'Why do we worry about description and why do we need to read up on how to do it?'

The basic reason is that the field exposure, be it a bag sample, a core or a trial pit is only seen (usually) by a single individual. This individual logger has to record all the information available from that exposure. This may or may not be straight forward but communication to those other parties who will use the information is a critical part of the exercise. The field log of the exposure is what the client who has commissioned the investigation is paying for, collection of essentially factual data on the ground conditions in order to inform the design process. The design process is one of engineering design and construction.

On the engineering side, many investigations commissioned by the design team concentrate on one or more aspects of the engineering problem, but most commonly on the foundations. However, most investigations actually require information on a number of other problem areas including those listed in Table 1.1 and illustrated in Figure 1.2 and Figure 1.3. For example any one borehole or trial pit may be addressing several of the aspects given in Table 1.1.

It is rare for an investigation to be as straightforward as a simple investigation for the building foundations. In order to construct the foundation, a slope or retaining

Table 1.1 Reasons for carrying out investigation

Investigation for a variety of engineering problems	Engineering aspects
Foundations	Strength – shear and frictional
Slopes	Compressibility
Retaining structures	Permeability
Bulk excavations	Fabric – affects strength and permeability
Ground treatment	Variability or inhomogeneity
Re-use of materials	
Contamination	
Groundwater (flow, level, quality)	
Construction failures	

Figure 1.2 Engineering works example – slope instability.

Figure 1.3 Engineering works example – piled foundations.

structure around the building footprint may be required in order to allow access for construction of footings and floor slabs. There could be need for bulk excavation to remove the ground to form basements. The information required by the foundation designer is very different from what the contractor needs to assess the ease of excavation and possible re-use of the arisings on site. Similarly, the piling contractor, the dewatering specialist and the geo-environmentalist all need different information on the ground in order to achieve an efficient project for the benefit of the client who is paying for the investigation.

The logger needs to balance all these potentially conflicting requirements in preparing a field log and communicating the observations to others. Good logging therefore clearly needs a practical understanding of the effect of the works on the ground and the effect of the ground on the works. This requirement conflicts with the situation where most logging is carried out by the more junior members of the industry who are unlikely to have gained an appreciation of these interactions. In addition, logging is generally carried out by geologically trained persons who, while they should understand the ground, often have little or no appreciation of the engineering aspects of the project.

The preparation of a description that is outlined in this book combines the use of a defined terminology with the English language to arrive at a standardised description. A system and a set of defined terms are necessary in order that any reader of the log, be they the client, the design office or the contractor preparing a tender, can correctly understand the ground conditions at the site that will affect any pertinent aspect of the construction of the works with which they are concerned.

In addition to this broad set of requirements that the logger must undertake, they must also be familiar with the full and wide range of soils and rocks that are encountered in investigations. This includes materials with shear strengths ranging from a few kPa, to rocks with a compressive strength of several hundred MPa. This wide range of materials arises from a variety of depositional and lithification processes over which is laid the effect of weathering processes.

The constituents of the ground that are to be described cover the wide range of materials outlined in Table 1.2. Examples of some of these materials are shown in Figure 1.4 which includes fine soil, coarse soil and rock both in core and hand specimens.

Engineering properties of the ground are governed by physical properties. It is usually possible to carry out tests to measure these properties, but tests are only carried out at particular depths or on individual samples. Interpretation of the test results requires some knowledge of the characteristics of the soil or rock being tested. In order to be able to interpolate these test results vertically within a borehole succession and laterally to other parts of the site, details of the intervening materials are needed. This means that description is a crucial piece of the jigsaw underpinning and linking together several aspects of the investigation.

In addition and perhaps most critically, when it is not possible to recover undisturbed samples or carry out in situ tests for whatever reason, the description is the only source of information on the properties of the ground. This sort of problem is most acute in hard soils, weak rocks and ground that is disturbed or made up.

The logging of field exposures is therefore an exercise in recording information that can be communicated to all parties to the proposed works, but this can only be

Table 1.2 Categories of materials

Class	Example	Typical feature
Soils	Soft alluvium	Low strength soils
	Peat	Compressible, possibly with high tensile strength
	Fine sands, silts	Soils that are sensitive owing to small but significant fines fractions and density
	Sandy gravels	Coarse permeable soils with variable proportions
	Overconsolidated clays	High material strength often reduced by fissuring
	Glacial tills	Wide range of particle sizes and strengths and high strength soils
Weathered rocks	Dense sand	Weak rocks/stiff soils that are transitional in behaviour and which are difficult to sample and test
	Stiff clay	
	Extremely weak mudstone	
Rocks	Mudstone	Weak rocks susceptible to degradation
	Sandstone	Weak to strong abrasive rocks
	Limestone	Strong rocks with voids
	Evaporites	Soluble rocks subject to creep deformation
	Granite	Very strong rocks
	Basalt	Weak to strong variable rocks
Made ground	Soils placed by man	Includes tipped materials, soils placed under engineering control and manufactured materials all contributing to huge variability and unpredictability
Fluids	Water or other fluids	Water as such does not need description but any observations relating to water, other fluids or gas, whether this is natural or some form of contamination must be described (e.g. soils being wet or dry, smells, hydrocarbon sheens, gaseous emissions)
	Gas	

Figure 1.4 Examples of the range of soils and rocks, in cores and hand specimens.

achieved if the logger follows a system that enables the description to be read and understood. The reader also needs to be aware of this system in order to extract the information he or she wants. This is why there is a systematic approach to description which all loggers and log users and readers can adopt. This systematic approach is enshrined in the Codes and Standards covered within the purview of this book.

There is also, however, a balance that needs to be drawn between a rigid systematic approach and a more flexible approach that allows and enhances the communication process. The logger needs to be aware of all this if they are to sift through the mass of information that could be recorded and only report those elements of the soil or rock that really do matter.

Over the years, there have been a number of cases where the logging geologist does not even give themselves a fighting chance of reaching these goals as they are not aware of the expected geology at the site. If the anticipation is not present, how can the actual findings be tested? Anyone carrying out logging in the field should make themselves aware of the likely geological succession and the proposed works. It does not take long to look at the geology map or a previous investigation carried out near the site. Is it really the case that the investigation team does not know why they are being paid to carry out the work?

However, if this is not possible, it remains the responsibility of the logger to record the factual evidence presented at the exposure. Interpretation of the geology starts with the description but the field log is also the point from which others can carry out more detailed and possibly a formation-specific interpretation of the geological succession and conditions.

The description of soils and rocks in engineering practice attempts to achieve a representation of the ground on a field record of some sort. The materials being described are put there by Nature (usually) and some humility and respect for our planet and the processes that made it are required. If investigators and loggers are arrogant and think they can beat Nature, they will probably trip up at some point, possibly quite badly. There have been too many collapses and other failures with subsequent forensic investigations have arisen because of inadequate ground description. The range of conditions that loggers are called on to describe is almost limitless and no one person will have seen all there is to see. It is true that the best logging geologist will be the one who has seen the most soils and rocks, but surprises are still possible even for the most experienced.

Field loggers should always be willing to ask others for help and should feel confident enough to ask for this help. To make such a request is an admission of competence, not a failure of knowledge on their part. Checking logging at the face or in the store where possible is ideal. If this is not possible, taking subsamples to the office is a means of obtaining the required help. It is particularly important to note that help is not just available from seniors. Good advice could also be available from peers, or even from juniors. The request for help should be addressed to the logger who is most experienced in describing the particular ground where the problem or difficulty has been encountered.

1.2 Description compared with classification

It is necessary at the outset to be clear about the difference between description and classification. This is neatly encapsulated in the introductions to the recent international standards. EN ISO 14688-1 establishes the basic principles for the general identification and description of soils and provides 'a flexible system for immediate (field) use by experienced persons'. Description is thus seen as the activity carried out in the field at the drilling rig, trial pit or core shed and does not require test results. It does however require that the logger is properly calibrated so that the descriptions are accurate. The matter of checking field logs against test results is discussed in Chapter 16.9.

The classification principles, outlined in EN ISO 14688-2, permit soils to be grouped into classes of similar composition and geotechnical properties, with respect to their suitability for geotechnical engineering purposes on the basis of test results. A number of national and international classification schemes have been proposed over the years and some examples are outlined and discussed in Chapter 15.

In principle, exactly the same distinction between description and classification applies to rocks, but EN ISO 14689-1 is presented in a single part, so that these two aspects are considered together within the Standards.

The application of classification systems to rocks has necessarily proceeded from a different starting point to that in soils. In soils the material properties are as important or more important than the discontinuities; in rock the converse is true with the number and style of discontinuities dominating the behaviour. The most common type of classification used in rock is the one of the mass rating systems such as the rock mass rating (RMR) or Q systems as discussed in Chapter 11.

The evolution of geotechnical practice in the UK followed previous practice in the USA before the 1950s as outlined by Terzaghi, Peck, Legget and others. This use of precedent practice applied particularly strongly to classification, in the use of the Unified Soil Classification System, for example. However, in description, UK practice evolved in a different direction and so current practices in the UK and USA differ.

1.3 Communication in description

Figure 1.5 is a real example of a sample description sheet prepared by a typical new graduate, that is, an individual who did not have any formal training in soil description on joining a site investigation company.

In these descriptions there are a number of undefined words used. What is the reader to make of the following unhelpful descriptions?

- Fine to medium soil – fine or medium soil in relation to what?
- Crumbly – does this mean it is friable or rather sandy?
- Large pebbles – what is the size of a pebble and what is a 'large' one?
- Coarse grains – are the coarse grains of pebble size or different?
- Dark black – is this a very unusual soil requiring further thought?
- Weak soil – how weak is a weak soil?
- Incoherent – this part of the description is at least correct.

Other examples of loose terminology met with on recent field logs on actual projects include:

- Loose claystones – as claystones are nodular concretions set in a mudrock matrix, what is loose?

Location	Leeds Road, Odsal	Location No	01/0368/A
Described by	ANO	Borehole No	42A
Date	08/01/2001	Sheet	1 of 1

Sample No	Type	Depth from	Depth to	Description
3	D	1.00	2.00	Very weak fine to medium SOIL. Very dark brown crumbly with large very rounded to subangular pebbles of various sizes. Incoherent
4	B	2.00	2.50	Weak locally quite firm fine to medium dark brown SOIL with subrounded to angular coarse grains. Clayey
5	D	3.00	3.50	Very weak fine to coarse brown to dark black SOIL with large gravel. Silty

Figure 1.5 Example of poor descriptive technique.

- Friable clayey SAND – is this a dry strong soil or is it just the sand grains that rub off?
- Very fissured, heavily fissured, slightly fissured – use of these terms gives NO idea of the fissure spacing which is so important.
- Heavily fractured, intensely fissured – use of undefined terms is not helpful because of subjectivity.
- Bluish orangey brown – the logger is obviously confused as to which Standard is in use with mixed terminology, which raises other worries.
- Olive clay, olive gravel – today, when our food comes from global sources, what sort of olive does the logger have in mind – black, green or some other colour?
- Very flinty GRAVEL – this begs the question of what other materials are present but not described.
- CLAY laminated with sand – it is not clear whether the clay is laminated or not, or if this means interlaminated materials. The outcomes are not the same.
- Verticle (*sic*) fissure, dip 10° – dip of fissure very uncertain.

The point at which these descriptions fail is that the reader has no possibility of unambiguously understanding what the materials are that the logger has seen. These descriptions would be examples of mission failure as communication of the characteristics of the sample or field exposure to the reader is not achieved.

In order to achieve the required level of clear and unambiguous communication it should be noted that the description of soils and rocks is both a science and an art. The scientific element of soil and rock description is the application of a systematic approach using a defined terminology for description, covered at length in this book. The systematic approach which underpins the whole subject area is critically important to the field logger, to ensure that all important aspects are covered and reported in a comprehensible way. It is also critical for the reader and user of the descriptions that arise from fieldwork to facilitate their extraction of relevant information from the description and application to the design question being addressed.

The artistic element of a description consists of ensuring that the reader of a description quickly and easily gains an accurate image of the ground seen by the logger and presented in the field description. This element includes the sensible use of the defined terminology and word orders, but incorporating the flexibility to ensure that the description is understandable in plain language. Concealment of the condition of a soil or rock in the ground by obscure terms or language or inappropriate levels of detail is helpful to no one, least of all the logger. If total clarity and a correct picture of the samples or exposures seen and described are not available to the user, the description fails and problems or claims may ensue.

1.4 Soil meets rock

The logging geologist is faced with soils which range in shear strength from 1 to 300 kPa and rocks which range in compressive strength from 0.6 to over 1000 MPa. In early codes, preceding BS 5930: 1999, the whole ethos and approach to description of soils and rocks came from different sources, as evidenced by the previous sentence which included different units and different measures of strength. The treatment of soils and rocks has historically been dealt with in different university departments (soil mechanics and rock mechanics) and this is part of the reason for the different approaches. For example, the strength of soil is reported as shear strength (undrained), c_u, and of rock as compressive strength (undrained), q_u which are respectively the radius and diameter of the Mohr circle.

This different source of the subjects gave rise to differences in approach to description. In particular, the boundary between soil and rock was not defined, with very stiff soils having a shear strength greater than 150 kPa with no upper limit. Conversely, very weak rocks were defined as having compressive strengths below 1.25 MPa, with no lower limit. Other differences included different word orders and little recognition that soils as well as rocks can be weathered and that this should be described.

In reality the transition between the stronger soils and the weaker rocks consists of a continuum either through varying levels of consolidation or lithification in a fresh material or different degrees of degradation in a weathered sequence. This latter case is shown in Figure 1.6 where there is clearly transported soil at the top of the exposure and weathered bedrock at the bottom. In between there could be more than one rockhead and the level of any actual boundary between the soil and the rock depends on why it matters.

This gradational range is important because this boundary area between soil and rock is exactly that area where sampling and testing are very difficult, if not impossible, as illustrated by the materials visible in Figure 1.6 and the criteria indicated in Table 1.3. The available methods of investigation and sampling of these materials tend to be compromises and description of the exposures often needs to fill the gap in test results.

The process of combining the description of the two types of material is now well advanced and the description of soils and rocks is genuinely a continuous spectrum

Figure 1.6 Exposure of gradational rockhead passing from bedrock at the base of the cliff to transported soil at the top. Various different rockhead levels could be defined within this gradation.

Table 1.3 Types of materials at the soil/rock boundary (after Norbury *et al.*, 1982)

Material type	Recognition	Investigation methods	Rockhead level	Excavation method
Overburden of transported soils	• Geological setting • From field descriptions • Gravel other than from underlying bedrock • Evidence for transport and re-deposition in the fabric	• Probing • Cable percussion • Trial pitting	Geological rockhead	• Scraping • Dozing
Weathered bedrock	• Angular gravel to cobble size fragments derived from underlying strata • Low TCR, SCR • RQD = 0 • If small or NI	• Refusal of cable percussion • Start of rotary coring	Engineering rockhead	• Dozing • Light ripping
Bedrock	• TCR high • SCR, RQD > 0 • If <80 mm	• Rotary coring		• Medium ripping
	• TCR=100% • SCR, RQD > 25% • If >70 mm	• Rotary coring		• Heavy ripping • Blasting

which reflects the range of engineering materials in this area. The description of soil and rocks now uses the same word order and terminology throughout, at least in UK practice, as discussed in Chapter 4.9.

The emphasis placed on the soil to rock transition might be thought to be a rather academic point but there have been a number of conferences and technical meetings addressing the subject of weak rocks over many years. The reason for the continued interest and difficulty in this area is exactly the same difficulty experienced in investigating and characterising such materials. The site investigation industry has considerable and ongoing difficulties in carrying out exploratory holes, in recovering representative samples and carrying out meaningful tests in the approximate range $c_u = 300$ kPa to $q_u = 1$ MPa. There have been a number of developments such as rotary triple tube drilling and in situ testing to assist with this but these difficulties remain in everyday commercial practice.

In those cases where representative samples may not have been recovered and if they have been recovered are not sensibly testable, the field description given by the logger is often the only information available. The description in these cases is all that the designer has to characterise the ground and to assign appropriate design parameters. The onus on the logger to identify the sequence correctly therefore increases at the very time that the most disturbed samples are generally presented from the borehole or trial pit.

1.5 Health and Safety in description

It is a general truth that field logging is carried out by young staff frequently working alone or at best in small numbers. For this reason alone, they deserve special care and attention to their wellbeing. Many others go out into the field with varying degrees of site experience and they also merit care to ensure that they are not harmed by the tasks and the work load that goes along with carrying out soil and rock description. Whilst there have seen very significant advances in the standard of risk assessment and safety briefing of field staff, the logging of samples, cores or excavations by graduate staff is probably the greatest hazard they are exposed to. It is therefore incumbent on all involved in site investigation to ensure that logging staff are adequately trained and briefed to enable them to carry out their work safely. This text is not intended to provide definitive guidance on health and safety matters in this area, rather the comments below result from years of working in this environment.

The main risks faced by logging personnel include:

- Manual handling in moving tube samples, bulk bags and core boxes between the field, sample store and transport and from storage racking to the description bench within the core store. Particular care is required when moving these items within the sample store during logging, as this is often carried out by the logger without the assistance of drilling crews or other labour. The provision of appropriate equipment for moving and lifting these loads is a very good way to limit possibilities of injury, as shown in Figure 1.7. Ensuring that the bags, tubes or boxes are sized appropriately to ensure they are not too heavy is also of great importance. It is better for a large bulk sample to comprise two bags of reasonable size rather than one overly heavy bag.
- Protection against contaminants, both anthropogenic and natural. The level of awareness of the risks posed by contamination is increasing all the time, but it is essential that the findings of the risk assessment are provided for the logging staff. In particular any pre-site briefing should bear in mind the relative inexperience of these staff. Further, the investigation is quite likely to uncover unexpected materials and risks, because it is an investigation and

Figure 1.7 Use of appropriate handling equipment in and around core and sample stores areas can reduce the risk of manual handling incidents and injury to loggers.

the safety briefing needs to include appropriate response procedures. Whilst it is thankfully unusual to dig up, say, medical waste in a trial pit without warning, it can happen and the field logger needs to have appropriate guidance to enable them to take the correct action. The provision of appropriate welfare facilities so that loggers can keep their hands clean and wash them before taking refreshment is also helpful.

- Work in and around trial pits and excavations requires that full risk assessments are made and in place. Even if man entry is not required, which is nowadays the norm, the logger needs to be alerted to the particular issues related to working with hydraulic excavators and to ensure that they do not fall into the pit or get struck by the excavator arm, just two of the most obvious causes of concern in pitting activities. Even standing near an open excavation can be hazardous if the pit were to collapse, as shown in Figure 1.8.
- If man entry to any excavation is required, appropriate support, dewatering, gas monitoring and escape procedures need to be in place as required.
- Investigation points involve penetration through areas which contain services of various sorts and the importance of identifying these in advance so as to avoid the consequences cannot be underestimated.
- Logging staff are often working alone and this poses particular risks. In these days of universal availability of mobile phones, communications are greatly improved and hugely easier than even a few years ago. The risks, however, remain and appropriate reporting in procedures and frequencies need to be established and followed. Particular care is required to identify reaction procedures for those, hopefully rare, occasions when the scheduled call in report is not made.
- Logging requires the use of tools ranging from blades and knives and spatulas to hammers. These should be used with due care and attention to the risks they pose, which largely boils down to common sense. Knives are banned from some sites as they are a very common cause of injury to hands and

Figure 1.8 Pit collapse immediately after removal of shoring.

knees. When power tools are used for cutting samples or liners, the relevant risk assessments should be made and appropriate operating procedures and controls put in place.

- The use of hammers, knives, power tools and acid means that there will be occasions when there are hazards due to an activity in progress. It is critical to adopt appropriate eye protection at these times. Hearing protection may also be required when power tools are in use.
- Loggers should be using dilute chemicals such as acid or dyes to test for certain minerals or lithologies. These should be kept in a safe place and clearly labelled so as to not be confused with harmless liquids.

The reader is referred to Chapter 17.5 where safety considerations specific to logging exposures in the field are considered.

2 History of Description in Codification

We have now, in the early 21st century, reached a position of international codification in the description of soils and rocks. There are many who are not aware of the benefits that this brings to the practice of engineering geology in all countries that follow the practices outlined. To do so allows their practitioners to work around the world and also allows them to receive and understand field descriptions and logs prepared by others. This is a happy state that has been reached in a remarkably short period. In order to understand fully this current position and how we got here, the history of this progress is outlined in the following sections. One of the reasons for doing this is because anyone using field logs prepared during this evolutionary period will need to be aware of the changing background and terminology during the past five decades.

2.1 Prior to 1970

The first publication that addressed the standardisation of soil and rock description was the Code of Practice on Site Investigation, BS CP 2001: 1957. This publication said little in detail about soil description and even less about rock description. It did introduce the concept of systematic description, although no standard terms or definitions were provided, nor any details of a standardised approach. The effect on practice was therefore modest although it has to be remembered that the national geotechnical industries were very small at the time and communication between individual practitioners was very much better than is possible in today's large and international industry. However, a very important impact on soil description was made in that the Code laid down the precept that descriptions should reflect the engineering behaviour of the soils; this should also be taken to apply to rocks. This precept still underlies the principles that are followed today making it clearly of fundamental importance.

Following that publication, individual companies still tended to carry out field logging in their own way as no standardisation of descriptive terminology was provided. As most logging in those days was carried out by geologists who came from a range of training backgrounds, descriptions tended to be individualistic, certainly in comparison with today's standards. Field log descriptions, generally of boreholes in those days, were not consistent or standard. On that basis, the criterion of descriptions being important tools for communication was not being met. For this reason the Engineering Group of the Geological Society (EGGS) set up a series of Working Parties (WP) to prepare guidance on standard nomenclature.

2.2 The period 1970–1981

The history of the codification of soil and rock description really starts in 1970 with the publication of the Engineering Group of the Geological Society (EGGS) Working Party (WP) report on the logging of Rock Cores (Anon, 1970). This pro-

Table 2.1 Summary of recommendations in Anon (1972)

Description item	Suggestion	Are proposals still in current practice?
Word order	No word order suggested.	Word order now provided
Colour	Terms provided; chroma terms have the suffix '-ish'.	Yes
Strength	Consistency terms for field use provided; strength classifiers provided.	Yes
Structure or fabric	Terms provided.	Yes
Weathering or alteration	Classification provided.	No
Particle shape	Three terms provided.	No
Composition	Record; limited guidance offered.	Yes
Grading	Terms provided.	Yes
Plasticity	Classifiers based on liquid limit.	Yes
Secondary constituents	Terms defined for coarse soils.	Yes
SOIL NAME	Terms provided; boundary between coarse and fine soil at 50%.	No
Permeability	Terms provided.	No

vided a first approach to systematic description and included the definition of terms for the description of colour, grain size, bedding and fracture spacing which we still use today. There have been some changes since those recommendations, particularly in the description of weathering and strength.

This publication was followed in 1972 by the EGGS WP report on the preparation of Maps and Plans (Anon, 1972). In addition to providing guidance on engineering geological mapping, this report provided additional guidance on description, in this case with terminology for use in describing soils. Thus, with the publication of this WP report, industry had been provided with the basis for systematic and standardised description of soils and rocks for the first time.

The guidance on soil description in Anon (1972) is remarkably long lasting, as shown by the comparison in Table 2.1. It can be seen that current practice is, in many regards, still in line with the recommendations made then.

Throughout the 1970s, work carried on to provide guidance in a number of different environments and this culminated in further publications. These reports are the EGGS report on Rock Masses (Anon, 1977), the International Society of Rock Mechanics Suggested Methods (ISRM, 1978, 1981) and the International Association of Engineering Geology Commission report (IAEG, 1981). These publications provided a range of proposals, with different coverage between the suggestions, as well as different proposals for the same item of description within the suggestions. This makes a confusing picture. The content of these suggested methods is summarised in Table 2.2.

The Working Party report on logging of Rock Masses (Anon, 1977) was also a step forward, but unfortunately introduced some inconsistencies from the earlier

Table 2.2 Summary of recommendations across the 1970 and 1980 series of proposed methods

	Rock cores report (Anon, 1970)	Proposal as current practice?	Maps and Plans report (Anon, 1972)	Proposal as current practice?	Rock masses report (Anon, 1977)	Proposal as current practice?	IAEG Commission report (1981)	Proposal as current practice?	ISRM Suggested methods (1978, 1981)	Proposal as current practice?
Word order	Implied with more important items coming first, with rock name last	No	No word order suggested		Description split explicitly into mass and material. Field description followed by classification of test results	Yes	Word order provided, starting with rock name, then material and mass characteristics	No	No preferred order given	No
Strength	Terms defined as used in UK until 2007	No	Terms defined as used in UK until 2007	No	Terms defined as used in UK until 2007	No	Terms defined, not used elsewhere	No	Terms defined	Yes
Colour	Included but no specific guidance provided	Yes	Terms provided. Chroma terms have the suffix '-ish'	Yes	Terms provided. Chroma terms have the suffix '-ish'	Yes	Terms provided. Chroma terms have the suffix '-ish'	Yes	No terms provided	
Bedding	Terms defined	Yes	Terms defined	Yes						
Grain size	Included but no specific guidance provided	Yes	Terms provided based on soil size fractions	Yes	Terms provided based on soil size fractions	Yes	Included and described without reliance on the rock name	No	No terms provided	

Table 2.2 (Continued)

	Rock cores report (Anon, 1970)	Proposal as current practice?	Maps and Plans report (Anon, 1972)	Proposal as current practice?	Rock masses report (Anon, 1977)	Proposal as current practice?	IAEG Commission report (1981)	Proposal as current practice?	ISRM Suggested methods (1978, 1981)	Proposal as current practice?
Grain shape							Terms provided as for soil	No		No
Texture and structure	Included but no specific guidance provided	Yes	Guidance provided including some suggested terms	Yes	Guidance provided including some suggested terms	Yes	Texture limited to grain size. Fabric terms to be included but not provided	Yes	No terms provided	
ROCK NAME	Included but no specific guidance provided	Yes	Included using simple terms that are correct	Yes	Classification for engineering purposes provided	Yes	Classification for engineering purposes provided	Yes	Classification for engineering purposes provided	Yes
Total core recovery (TCR)	Defined	Yes							Defined	Yes
Solid core recovery (SCR)	Full diameter definition	Yes								
Rock quality designation (RQD)	Defined using 100 mm	Yes					Record and loosely defined	Yes	Full diameter definition	Yes
Fracture index	Natural fractures only, over arbitrary length	Yes							Natural fractures only, over arbitrary length	Yes

Parameter	After Ege (1968)				
Stability index	No	No	No	No	See Chapter 2.4
Permeability		Terms provided based on jointing	Guidance provided. Testing recommended	See Chapter 2.4	No
Weathering or alteration	Classification based on discolouration and friability	See Chapter 2.4	Classification A. Use alteration terms where appropriate	Classification B. Terms given separately for material and mass	Material weathering classified on degree of change (%)
	See Chapter 2.4				Classification B. Terms given separately for material and mass
Discontinuity type	Record type	Yes	Record type. Short list of types given	Record. Guidance on types provided	Yes
					Record. Guidance on types provided
					Record. Definitions of types provided
Discontinuity orientation	Record but no guidance offered	Yes	Record but no guidance offered	Record. Guidance provided	Yes
					Record. Guidance provided
					Detailed guidance provided including methods of presentation of data
Discontinuity spacing or frequency	Terms defined	Yes	Terms defined	Terms defined	Yes
					Terms defined
					Terms defined
Discontinuity aperture	Record but no guidance offered		Terms defined	Terms defined	No
					Terms defined (terms differ from Anon, 1977)
					Terms defined
Discontinuity infilling	Record but no guidance offered	Yes	Record. Guidance provided	Record. Guidance provided	Yes
					Record. Guidance provided

Table 2.2 *(Continued)*

	Rock cores report (Anon, 1970)	Proposal as current practice?	Maps and Plans report (Anon, 1972)	Proposal as current practice?	Rock masses report (Anon, 1977)	Proposal as current practice?	IAEG Commission report (1981)	Proposal as current practice?	ISRM Suggested methods (1978, 1981)	Proposal as current practice?
Discontinuity persistence or extent					Record. Guidance provided	Yes	Measure and report	Yes	Measure and report. Terms defined	No
Discontinuity surface form	Record but no guidance offered		Record but no guidance offered		Record. Guidance and terms provided	Yes, not terms	Record. Guidance and terms provided	No	Terms defined	Yes
Discontinuity wall strength							Describe or measure and report	Yes	Describe or measure and report	Yes
3D arrangement of discontinuities			Terms offered without definition	Yes	Terms offered with definitions	Yes	Terms offered with definitions	Yes	Terms offered with definitions	No
Seepage							Estimate rate of flow	Yes	Terms defined	No

reports on core logging and mapping (Anon, 1970, 1972). The main inconsistency was the scale for weathering or alteration. The scheme presented was the same as that subsequently included in BS 5930: 1981 and this is discussed in Chapter 10. This report related to the description of rock masses and included the results of a number of laboratory and field tests such as permeability, modulus and density in the context of indices to describe the mass behaviour. This approach is no longer followed. The field description excludes test results which are considered separately in the ground assessment as outlined by the principles clearly stated in EN ISO 14688. The report (Anon, 1977) clearly called for distinction between description of the rock material and rock mass features and emphasised that the description of both was required and important. The report also called for any arbitrary distinction between soil and rock to be disregarded. The description of a mass was envisaged to apply equally to a jointed granite, a cleaved slate, a fissured London clay and a jointed glacial till. This sentiment is fully endorsed as the principles have underlain the evolution of the Standards. This ethos is also reflected in the content and structure of this book.

The proposed methods summarised in Table 2.2 were in preparation at the same time as BS 5930, which was itself published in 1981. It is interesting to consider what the status of these various suggestions for practice were. The proposals by EGGS are merely Working Party reports published by the Geological Society of London. However, their suggestions were widely adopted with much of the content being picked up in the ISRM and IAEG suggested methods. These were documents prepared by international working parties and so may have had greater status. Although this is not a legal position, these recommendations were being put forward as good practice by the industry and should have been followed. When BS 5930 was published this was at least a national standard, albeit published as a Code of Practice. Nevertheless these various documents were adopted in parts of the world, to a degree because of the absence of anything else. The problem is that the documents are not entirely consistent with one another and there are a range of approaches, terms and definitions provided, as shown in Table 2.2. Although the situation was improved by these documents, as there was no standardisation before, there is still ample room for confusion and ambiguity.

There is an explicit requirement in Eurocode 7 for the standards followed in the carrying out of any work to be identified within the report. The situation outlined above is a classic illustration of just how eminently sensible is this requirement. This practice is strongly commended to all – practitioners should ensure that all reports are very clear about the descriptive terminology standard used. This clarity should preferably accompany the borehole or field log sheets, rather than just being included in the report text. It is the general situation that the field borehole, trial pit or exposure logs tend to get separated from the rest of the report and particularly the report text.

2.3 The period 1981–1999 and the first BS 5930

The 1981 UK Code (BS 5930: 1981) was ground breaking as it was the first national code which included guidance on the description of soils and rocks. As such it was a major achievement, perhaps reflecting the 14 years it took to write. The fact that the authors stuck their necks out and tried to codify this area was brave, given that

BACKGROUND ON STATUS OF BS5930

The question of the status of BS 5930 has concerned engineering geological and engineering practitioners ever since its publication and the question of how practitioners are required to deal with 'should' guidance and 'shall' guidance still remains.

When BS 5930 was first published, PP 444 (Anon, undated) explicitly cited BS 5930 as a Code of Practice as follows:

Broadly speaking, two types of code may be distinguished. The first type details professional knowledge of practice, for example BS 5930 Code of Practice for site investigations. Normally they would not be called up directly in a job specification, although they could be mentioned in a preamble requiring the contractor to have a general knowledge of them. The second type is more specific, for example CP 118, which could be called up directly in a job specification using such words as 'The aluminium structure shall comply with...

It was clearly considered by BSI at the time of publication that a Code of Practice indicated professional practice which would not normally be called up directly in specifications.

until this time practising geologists had not undergone national control in this way, notwithstanding the various published proposals.

Despite the apparently intended status of BS 5930, practitioners could be expected to be aware of the recommendations. In reality, standard practice has become that BS 5930 is usually called up in Specifications and this elevates the recommendations in the Code to a contractual level and, with continued such usage, to standard practice. Similarly, a professional practitioner failing to follow this or any other Code of Practice would face difficulty in defending a negligence claim. The Code of Practice had therefore become a de facto Standard.

Given that this BS 5930: 1981 was a newly drafted Code, not all of the recommendations therein were found to be acceptable by the industry. This is hardly surprising and criticisms and suggestions for improvement were put forward at the Guildford Conference in 1984 whose proceedings were published in 1986.

Norbury *et al.* (1986) raised a number of issues on soil and rock description, in particular:

- The distinction between fine soil and coarse soil solely on grading was rejected in favour of description according to engineering behaviour as laid down in BS CP 2001: 1957. This issue is discussed in detail in Chapter 4.6.
- The distinction between silt and clay solely on the basis of the A-line was supplemented by additional terms for borderline cases. The varying practice in this area is discussed in detail in Chapter 4.7.
- Simplification and extension of terminology for the description of secondary constituents of mixed soils using the terms with 'a little'/'some'/'much', which could be used after the principal type, allowed the inclusion of detail about secondary constituents.
- Proposed weathering classification for rocks was rejected as inapplicable, in favour of the then current practice which was to use the proposals in Anon (1972). Concerns over the earlier proposals were considered in the

rock weathering working party (Anon, 1995), as discussed in Section 2.4 and Chapter 10.

- Rock nomenclature was clarified to conform to geological scientific conventions. The change here was to remove the specific grain size definitions for igneous and metamorphic rock as this correlation conflicts with standard geological nomenclature.

BACKGROUND ON DESCRIPTORS THAT ARE NO LONGER IN USE

The terms 'with a little'/'with some'/'with much' were used after the principal type terms and equate to the terms 'slightly'/'·' (the term)/'very', e.g. 'slightly sandy'/'sandy' (the term)/'very sandy'. These post-principal terms were dropped from use with the publication of the 1999 Code and the approach was replaced by the current multiple sentence approach.

These comments and suggestions variously found their way into national practice. While some companies were explicit in the approaches followed, most were not so rigorous. It is not clear how many practitioners actually read national standards such as these and how many simply follow the summaries and suggestions made in the technical press or by a limited number of colleagues. Nevertheless, it is apparent that the situation had been improved, but not resolved, by the publication of the BS. Only now, 20 years later, are we reaching a situation where the suggestions made in Standards are becoming accepted into mainline practice and falling explicitly under the remit of normative standards which are compulsory.

The practices recommended in BS 5930: 1981 were broadly adopted in other parts of the world, as they represented a benchmark of good practice in description. One good example of this is Geoguide 3 produced by the Geotechnical Control Office of the Hong Kong Government (Anon, 1988). The adoption here and elsewhere was qualified in a manner similar to the reservations expressed at the 1984 Guildford conference with additional variations to cover the particular geological conditions in those other parts of the world. Nevertheless the fact that the practice was being adopted widely was a significant step forward in practice generally.

The Tropical Soils Working Party Report (Anon, 1990) included guidance on the description and classification of such soils. The basis of the terms and approaches used was firmly based on earlier working party reports, continuing the sequence of evolving guidance documents, although there were considerable extensions into other areas, primarily of classification. For example, descriptors were provided for parameters such as moisture state, fabric orientation and durability as shown in Table 2.3 and Table 2.4. These terms could be useful on occasion in logging other materials.

Table 2.3 Description of moisture state (intermediate categories omitted)

Term	Definition
Dry	Lighter in colour than when moist. Sands are friable.
Moist	Exhibit range of colour change. Absence of wet or dry characteristics.
Wet	Water films visible on grains and peds.

Table 2.4 Fabric orientation

Orientation	Definition
Strong orientation	>60% of the particles are oriented with their principal axes within 30° of each other.
Moderate orientation	40–60% of the particles are oriented with their principal axes within 30° of each other.
Weak orientation	20–40% of the particles are oriented with their principal axes within 30° of each other.
No orientation	No preferred orientation.

2.4 Rock weathering

In the 1970s and 1980s there were two systems available for the 'description' of rock weathering. The first was that set out in WP Rock Cores (Anon, 1970) and was extended in WP Maps and Plans (Anon, 1972). The second had evolved outside the UK which tended to reflect the different effects of tropical and humid climate weathering, particularly on granitic rocks. These non-UK systems became the recommendations within BS 5930 (1981), ISRM (1978) and IAEG (1981).

These two approaches are very different in their view of how weathering takes place and they are contrasted in Table 2.5 and Figure 2.1 and Figure 2.2. This is not a semantic point as fundamentally different geological processes are being incorporated. It will also take only a few moments thought when reading this table to realise that the weathering processes envisaged within the first two columns are not commonly found in northern European countries and neither are therefore appropriate for use in these and in many other parts of the world.

The frequency with which construction claims, usually earthworks or tunnelling, were fought on the grounds of weathering classifications in the 1970s and

Table 2.5 Summary of weathering processes

Anon (1972) Classification A, Table 2.2	BS 5930: 1981, IAEG (1981) Classification B, Table 2.2	Anon (1995) and BS 5930: 1999
Weathering of the rock material throughout the rock mass is progressive, penetrating inwards from the discontinuities. A single set of descriptive terms is defined which relate to the degree of weathering of the rock mass as a whole.	Weathering proceeds into the fresh rock mass from the discontinuities more or less as a weathering front. There can be considerable changes in the degree of weathering of the rock material over small distances. Two sets of descriptive terms are defined, namely: • terms relating to the degree of weathering of the rock material • terms relating to the extent of (complete) weathering of the rock mass	The results of weathering of the rock are to be fully described. This will normally require description of changes in colour, changes in fracture state and changes in strength. The presence and character of weathering products is also to be described. Classification is only to be applied if useful and unambiguous.

Figure 2.1 Corestones of relatively fresh rock material within completely weathered material in Hong Kong as envisaged by the column 2 approach in Table 2.5 (photograph courtesy of Steve Hencher).

Figure 2.2 Interbedded sandstones and mudstones in Spain undergoing weathering as envisaged by the column 3 approach in Table 2.5 (photograph courtesy of Steve Hencher).

1980s suggests that engineering geologists were not delivering a reliable or satisfactory product to their client. This is hardly surprising given the above comments.

Notwithstanding these differences, it is important to realise that these systems for the description of weathering were not in fact descriptive schemes at all. These systems, which are still extant in EN ISO 14689-1, are in fact classifications for field use. This is nonsense and there are a number of reasons why such an approach using these systems is unsatisfactory and this can be seen in the examples in Figure 2.1 and Figure 2.2.

First, the use of a classification scheme rather than a description system as a primary tool in the field is not sensible (*q.v.* the explicit separation in soil description in EN ISO 14688 between field description and subsequent classification). Field logging must set out to record the features seen without being proscribed by specific classifications.

Second, the EN, ISRM and IAEG proposals assume that a single classification scheme along the lines of the proposals published will suit all rock types and all weathering processes. A moment's thought about the sheer variety of lithologies encountered in investigations and on which Nature has superimposed a variety of weathering processes, both historic and current, and the problem becomes clear. Loggers need to have a more flexible system available for use. Otherwise users of field logs will receive inappropriate information and thus struggle to achieve the hoped for levels of communication.

When BS 5930: 1981 was published, it was considered by many that the approach recommended was not suitable for most UK and northern European rocks as these have been subjected mainly only to temperate weathering mostly since the last major glaciations. It was therefore proposed (Norbury *et al.*, 1986) that the EGGS Working Party (Anon, 1970; Anon, 1972) combined schemes, whilst not perfect, still offered a better approach for the rocks that are generally encountered in northern Europe than that suggested in BS 5930: 1981. They recommended a reversion to the earlier proposals (op. cit.) so that it was the case that some of UK industry (not all) was using systems that contradicted the Standard. This was clearly an unsatisfactory position that was not tenable in the long run.

Given all the problems with existing classification schemes masquerading as descriptive approaches and the limited relevance to many rocks, EGGS decided in about 1990 to set up a working party to look at this in more detail. The main recommendation in the working party report (Anon, 1995) was that the first and mandatory approach to the consideration of logging all weathered profiles was actually to describe the weathering features that are seen in the rock, or indeed in the soil if that is where the weathering processes are visible.

This sounds a revolutionary step and in some ways it was, but in fact it was a reversion to earlier practice. Most logging geologists following the earlier classification recommendation would recognise weathering profiles by identifying a series of strata or zones through the weathering profile in which the strength, colour and fracture spacing differ. So the explicit requirement to describe was merely encouraging loggers to think about the processes that were giving rise to the stratification or other variation caused by weathering that they were observing in the field and already recording on their logs.

Having described the visible weathering effects in the core, the logger following the recommendations in Anon (1995) is then invited to consider whether classification of the weathering profile would be useful and whether there is a scheme available (or that could be established) that would be helpful and unambiguous. If the answer to these questions is in the negative, classification is not considered advantageous and should not be applied. This was a difficult step for many recipients of logs to accept at the time, as they had become used to receiving a weathering classifier on field logs. This was clearly a misguided requirement on their part.

If, on the other hand, a suitable classification scheme is available and useful, then it could be used. In the working party report, a variety of approaches are suggested for use. The route to follow in applying the general classification approaches proposed depends on the strength of the rock. The approaches proposed comprise:

- Approach 2 for rock materials that are medium strong or stronger (Note that the strength terms used in Anon (1995) have been superseded, as shown on Table 5.8.)

- Approach 3 for heterogeneous rock masses that are medium strong or stronger
- Approach 4 for homogenous rock materials and masses together which are weak or weaker
- Approach 5 is the any other business category for other rock types or weathering processes. Into this would therefore fall classifications of process or rock types such as karstic weathering (e.g. Waltham and Fookes, 2003), tropical weathering (Anon, 1990) or chalk (Lord *et al.*, 2002).

Within each of these broad approaches, a number of rock type, strength or climatic conditions can be placed. This style of approach therefore allows any classification that is applied to take some account of the rock types present and the weathering processes undergone. This was therefore a significant step forward and on this basis the recommendations in Anon (1995) were incorporated into BS 5930: 1999.

The procedure encompassed within Approaches 2 and 3 would be familiar to geologists working in humid tropical weathering environments. New procedures compatible with the recommendations of Anon (1995) have been developed for local conditions in countries such as Malaysia (Mohamed *et al.*, 2006).

The availability and application of the descriptive and classification approaches is given in more detail in Chapter 10.

2.5 1999 and the second BS 5930

The criticisms of the 1981 version of the BS 5930 had potentially compromised professional practice since publication. Because some of the proposals were not practical or workable, industry had to find means of applying the Standard in practice. The solutions adopted were not consistent and so some areas of confusion were allowed to remain in practice. These issues were addressed by introducing some significant changes in the 1999 Code revision. These included:

- The terminology for describing fine soils introduced in the 1999 Code was a further attempt to clarify and simplify the terminology for the description of fine soils. This was therefore yet another change in this difficult area for description. The changes are detailed in Chapter 4.7 below.
- In the consultation period, industry respondents suggested the use of multiple sentences in descriptions to make them easier to comprehend. This very sensible suggestion was taken up at a late stage and has achieved its aim.
- Terms for secondary soil fractions were changed with the omission of the 'with a little'/'some'/'much' terms that had been suggested by Norbury *et al.* (1986).
- Reporting of relative densities came to a position where it was only to be applied with N values (where N is the blow count from the standard penetration test). This removed the terms previously used in trial pits which related to the ease of driving a 50 mm peg. Although this is a reasonable approach in Codes, there is still a need for these simple field tests and this is discussed further in Chapter 5.1.
- The word order for rocks was adjusted so that the word order for soils and rocks was the same. The reason for this was to make the description of material at the soil–rock boundary easier, thus allowing descriptions such as very stiff CLAY/extremely weak MUDSTONE or very dense SAND/extremely weak SANDSTONE to appear.

> **TIP ON WORD ORDER AT THE SOIL–ROCK BOUNDARY**
> - The standard word order for soils and rocks has been made the same.
> - The reason for this was to make the description of material at the soil–rock boundary easier, allowing descriptions such as 'very stiff CLAY/extremely weak MUDSTONE' or 'very dense SAND/extremely weak SANDSTONE' to be used.

- Description of rock weathering was required and this was to be followed by classification only where this is available, useful and unambiguous as out-lined above.

2.6 The period since 1999

The story of codification comes up to date with the most recent international stand-ardisation contained within EN ISO 14688 and EN ISO 14689 published between 2002 and 2005. These Standards are called up as normative procedures in EN 1997-2 (Eurocode 7) and are remarkable steps forward for two reasons.

First, these are no longer guidelines or suggestions provided by an individ-ual author or the working party of a national or international learned society or in a national code of practice, but normative rules contained in a Standard. Any uncertainty about the guidance provided in comparison to the previous codes is removed.

However, readers of the new Standards will note the inclusion of a number of 'should' statements even in these normative soil and rock description standards. The status of these is uncertain, similar to the 'should' statements in BS 5930. It is suggested that all practitioners should err on the side of caution and assume these statements are closer to 'shall' statements, at least for the present. This considera-tion has the same outcome as discussed in Section 2.3.

The second step forward is that these are now international Standards. They were prepared under the aegis of the International Standards Organisation (ISO). In accordance with the Vienna Agreement the draft Standards were circulated and voted on in parallel by member states of both ISO and CEN and hence the joint EN and ISO identifiers in the Standard reference. These standards have therefore been adopted not only throughout Europe but also internationally through all member states of CEN and ISO, respectively.

That we have gone from there being little or no national standardisation in 1970, to national, European and now even international standardisation in a period of only 37 years is an outstanding achievement. This step alone should, in theory, go a long way toward improving mobility of professionals, one of the original aims of establishing the Eurocode system.

The changes to UK practice introduced in EN ISO 14688-1, EN ISO 14688-2 and in EN ISO 14689-1 were discussed in detail by an *ad hoc* industry working group (Baldwin *et al.*, 2007). This paper provides a clause by clause comparison of the new Standards against existing national practice. The two approaches do not map perfectly across one another and so certain recommendations needed to be made to allow implementation into practice. These are taken to be local enhance-ments to practice which are necessary to remove residual ambiguities. They are

BACKGROUND TO AMENDMENTS AND THEIR STATUS

- An Amendment to a national Standard is where the changes are the minimum needed to implement the EN ISO Standards. The key component of an Amendment is that it involves a lower requirement on Public Comment.
- The BSI numbering system is such that an Amendment does not change the title or date of the original reference. Thus the 2007 amended document is referenced as BS 5930:1999 Amendment 1.
- This change in reference is subtle, being presented in small font on the front cover and in the document version tracking. The important point is that any revision or amendment automatically overwrites the predecessor document. Thus any reference to BS 5930:1999 is, by default, to Amendment 1.
- If you are trying to state that you have carried out logging to the prior version of the Code, the reference will have to be 'BS 5930:1999 without Amendment 1' or 'pre-Amendment 1'.
- BSI rules are such that only two Amendments to a Standard are permitted. Thereafter, the Standard would have to be revised. The Revision process allows changes to be made throughout a standard (for reasons of correction, update or whatever).
- A revision would include a revised publication date.

certainly not areas of disagreement or conflict and so this approach is reasonable and legitimate.

The recommendations in Baldwin *et al.* (2007) were incorporated into amendments to BS 5930 in 2007 and 2010. The changes within this document were solely those in Section 6 relating to the description and classification of soils and rocks. Further amendment and revision of the BS will follow.

2.7 Multiple usage of defined terms

The significant steps taken in codification over the last few decades come at a price, however. In the various documents that have been published over the years the same terms have in a number of instances been used repeatedly but with a range of definitions. The most significant of these are summarised below.

2.7.1 Clay and silt terminology

The use of the terms 'clay' and 'silt' as principal and secondary constituents in fine soils has varied, in particular as to whether the terms are exclusive or not and how mixtures of fine soils are to be described.

Before 1981, general practice was for most fine soils to be described as 'silty CLAY', with a few marginal cases indicating higher silt content (or lower plasticity) by using the adjective 'very silty' before the principal 'CLAY'. Many clays were erroneously described as silty merely because of the presence of shiny flecks of mica or shell fragments. This practice was not helpful nor based on any genuine observations during most logging. This is one of the main reasons why this has been discouraged in engineering geological practice. The practice of describing most fine soils as 'silty CLAY' remains widespread in many parts of the world and is also the norm in standard geological usage.

A large majority of natural fine soils do, in practice, include both clay and silt fractions. Indeed pure clay or pure silt soils are reasonably uncommon, although they do occur as bentonite soils or rock flour, for example. Everyday practice in the description of transported soils can safely assume that the fine soil component is a mixture of both silt and clay. The question is how to assess the relative proportions or influence of these two size fractions.

Because of the poor practice in engineering geological logging, the answer to this has normally been to consider the two size fractions as exclusive. The fine soil is thus either 'CLAY' or 'SILT'. This has been the approach adopted in BS 5930. The 1981 version implied such an approach but unfortunately diluted the clarity by giving combined description terms such as 'silty CLAY' in examples. In the 1999 revision, this concept was stated explicitly and conflicting examples were removed.

Practical guidance has considered and evolved a variety of ways of identifying and reporting the range from CLAY to SILT and the intervening boundary based on the behaviour of the soil. The 1981 Code made the following suggestions:

> The soil name is based on particle size distribution and plastic properties. These characteristics are used because they can be measured directly with reasonable precision and estimated with sufficient accuracy for descriptive purposes. They give a general indication of the probable engineering characteristics of the soil at any particular moisture content.
>
> —Clause 41.3.2.1

Unfortunately, the normal sedimentation tests do not reliably measure the grain size in fine soils. Norbury *et al.* (1986) proposed the terms 'very silty CLAY' or 'very clayey SILT' to identify borderline fine soil materials. This was an attempt to fit closely with the 1981 Code and was reasonably successful. However, it appears that there was an unforeseen outcome in that the proportion of SILT in the natural world apparently increased during this period.

The stricture in Norbury *et al.* (1986) and BS 5930:1999 was that the use of the hybrid terms flagged that the soil was borderline between CLAY and SILT. If this borderline position was important to the engineering aspect of the works under investigation, then further sampling and index testing should be carried out to resolve the uncertainty flagged by the logger through the use of such terms. This approach generally worked well, although many geotechnical engineers still appear reluctant to carry out the sort of comprehensive programme of index testing envisaged by such statements.

This approach was modified in the light of experience such that the 1999 Code replaced these terms for the borderline case with the hybrid term 'CLAY/SILT'. In order to avoid any ambiguity or attempts at being clever, the Code explicitly stated that this term was synonymous with the alternative term 'SILT/CLAY' leaving just a single borderline term.

In particular Clause 41.4.4.1 identified that 'all soils should be described in terms of their likely engineering behaviour, the descriptions being supplemented with and checked against laboratory tests as required' and in Clause 41.4.4.4 'the distinction between clay and silt is often taken to be the A-line on the plasticity chart, with clays plotting above and silts below; however, the reliability of the A-line in this regard is poor, as might be expected'.

Table 2.6 Summary of terms for description of borderline fine soils

Standard	Thoroughgoing clay	Materials borderline between clay and silt	Thoroughgoing silt
Pre-1981 Code	Silty CLAY	Very silty CLAY or Very clayey SILT	Clayey SILT
1981 Code	CLAY	No terms	SILT
Norbury *et al.* 1986	CLAY	Very silty CLAY or Very clayey SILT	SILT
1999 Code	CLAY	CLAY/SILT (single term)	SILT
2007 Standard	CLAY	Silty CLAY or Clayey SILT	SILT

However, the rules given in EN ISO 14688 take us backwards as they reintroduce the terms 'silty CLAY' and 'clayey SILT'. Because this is a normative Standard in Europe, this rule has to be followed in order to accord with the rule of supercession. Not to follow this rule and stay with the above hybrid terms would produce a national conflict with the European standard and this is not allowed. The solution to this difficulty has been to replace the term 'CLAY/SILT' with the terms 'silty CLAY' or 'clayey SILT', that is to change the name used for the borderline materials. This is not entirely satisfactory, but it is hoped that this will continue to provide a workable methodology.

The position on the descriptive terminology for fine soils has therefore clearly suffered a number of changes before arriving at the current practice. This sequence of changes is summarised in Table 2.6.

Procedures for the identification of clay–silt mixtures and the borderline materials in field logging are described in detail in Chapter 4.7.

2.7.2 Secondary constituent terms

Terms used to indicate the proportion of secondary constituents with a 'little'/'some'/'much' were proposed by Norbury *et al.* (1986) for all secondary constituents. These terms are now not used in this way, having been replaced by the multiple sentence approach (see Chapter 3.2). The terms are still defined for use only in the description of some mixtures of very coarse soils.

2.7.3 Loose and dense

A variety of terms have been used to classify relative density on field logs. A variety of terms have also been used to describe the density or condition of a range of size fractions, with the same terms being used more than once. These various definitions are presented on Table 2.7 and it should be noted that the terms on the right hand side of the table are those that are no longer included in the permissible vocabulary. The correct use of these terms is described in Chapter 5.1 where some possible additional terms for use in the field when N values are not available are also discussed.

2.7.4 Compactness of silt

A variety of terms have been used to describe the consistency of silt. These vary from loose or dense in BS 5930:1981 (see Table 2.7) to compact and uncompact in

Table 2.7 Density terms – current and superseded shown shaded

Permissible terms			Non-permissible terms (previously used)		
SPT N value N_{60}	Sands and Gravels in boreholes	SPT value $N_{1(60)}$	Sands and gravels in trial pits	Silts (by hand test)	Cobbles and boulders (by inspection)
0–4	Very loose	0–3	Loose	Loose	Loose
4–10	Loose	3–8			
10–30	Medium dense	8–25	Loose or dense?	Dense	
30–50	Dense	25–42	Dense		Dense
Over 50	Very dense	Over 42			

BS 5930:1999. The permissible terms have been changed again in EN ISO 14688-1 where the same terms are used as for clay. Whilst this may not be a safe idea, it is at least straightforward and clear. The appropriateness of these terms is discussed in Chapter 5.2.

2.7.5 Rock strength

UK practice has been to follow the terms in Anon (1970) and Anon (1972) and this has remained consistent until now. In the meantime, ISRM, and thus broader international practice, has used different classification definitions but again with similar terms. Thus to follow the definitions included within EN ISO 14689-1 requires that the UK fall in line with practice elsewhere. This is a long overdue move to compatibility with international practice but requires significant changes to UK practice, as discussed in Chapter 5.3 and readers familiar with UK practice in this area are encouraged to refer to the changes that are presented below.

2.7.6 Rock weathering

Terms have been suggested in a variety of definitions, as discussed in Section 2.4 above. The terms have been perpetuated variously and are widely and generally considered as descriptors. Along with the various different approaches, the same terms have been used to report on very different definitions of classes of weathering. For example, the terms 'slightly', 'moderately' and 'highly' weathered have been defined variously as indicated in Table 2.8.

The historical issues concerning weathering description and classification have been discussed in Section 2.4. The current approach to and terminology for use in weathering description and classification where appropriate, is given in Chapter 10.

2.7.7 Comparison of descriptive terminology

Despite the many and varied changes that have taken place in the description systems over the years, there are actually fewer changes than many practitioners believe. In order to illustrate this example descriptions for a range of materials are given in Table 2.9.

Table 2.8 Comparison of weathering descriptions

Standard	Slightly weathered	Moderately weathered	Highly weathered
Anon (1970)	Penetrative weathering has developed on open discontinuity surfaces but there is only slight weathering of rock material.	Weathering extends throughout rock mass but the rock material is not friable.	Weathering extends throughout rock mass but the rock material is partly friable.
Anon (1972)	Rock may be slightly discoloured particularly adjacent to discontinuities which may be open and have slightly discoloured surfaces; intact rock is not noticeably weaker than fresh rock.	Rock is discoloured, discontinuities may be open and will have discoloured surfaces with alteration starting to penetrate inwards; intact rock is noticeably weaker than the fresh rock.	Rock is discoloured; discontinuities may be open and have discoloured surfaces; original fabric is altered near discontinuities; alteration penetrates deeply but corestones are still present.
IAEG (1981)	Degree of weathering change is up to 10%.	Degree of weathering change is 10–35%.	Degree of weathering change is 35–75%.
ISRM (1980), BS 5930 (1981), EN ISO 14689-1 (2004)	Discolouration indicates weathering of rock material and discontinuity surfaces; all the rock material may be discoloured.	Less than half of the rock material is decomposed or disintegrated to a soil. Fresh or discoloured rock is present as a continuous framework or as corestones.	More than half of the rock material is decomposed or disintegrated to a soil. Fresh or discoloured rock is present as a discontinuous framework or as corestones.
Anon (1995), BS 5930: 1999 pre Amendment 1	Slight discolouration, weakening	Considerably weakened, penetrative discolouration	Does not readily slake when dry sample immersed in water.

Table 2.9 Descriptions from 1981 to 2007 compared

BS 5930: 1981 description amended by Norbury *et al.* (1986)	BS 5930: 1999 description	EN ISO 14688 & 14689, BS 5930: 1999 Amendment 1 (2007)
COARSE SOILS		
Loose orange brown fine to coarse SAND with some fine and medium rounded gravel of quartz (RIVER TERRACE DEPOSIT)	Loose orange brown gravelly fine to coarse SAND. Gravel is rounded fine and medium of quartz (RIVER TERRACE DEPOSIT)	Loose orangish brown gravelly fine to coarse SAND. Gravel is rounded fine and medium of quartz (RIVER TERRACE DEPOSIT)
Medium dense brown grey slightly clayey subrounded fine GRAVEL of medium grained sandstone with some medium to coarse sand (GLACIAL TILL)	Medium dense brown grey slightly clayey medium to coarse sandy subrounded fine GRAVEL of medium grained sandstone (GLACIAL TILL)	Medium dense brownish grey medium to coarse sandy slightly clayey subrounded fine GRAVEL of medium grained sandstone (GLACIAL TILL)
Dense brown subrounded fine to coarse GRAVEL of various lithologies with a little sand and occasional subrounded cobbles of strong sandstone (FLUVIO GLACIAL DEPOSIT)	Dense brown slightly sandy subrounded fine to coarse GRAVEL of various lithologies with occasional subrounded cobbles of strong sandstone (FLUVIO GLACIAL DEPOSIT)	Dense brown slightly sandy subrounded fine to coarse GRAVEL of various lithologies with low cobble content. Cobbles are subrounded of strong sandstone (FLUVIO GLACIAL DEPOSIT)

Table 2.9 *(Continued)*

BS 5930: 1981 description amended by Norbury *et al.* (1986)	BS 5930: 1999 description	EN ISO 14688 & 14689, BS 5930: 1999 Amendment 1 (2007)
Grey and light grey subangular tabular fine and medium weak mudstone GRAVEL with much medium to coarse sand (RIVER CHANNEL/WEATHERED BEDROCK?)	Grey and light grey very medium to coarse sandy subangular tabular fine and medium weak mudstone GRAVEL *or* Very sandy GRAVEL. Gravel is grey subangular tabular fine and medium of weak mudstone. Sand is light grey medium to coarse (RIVER CHANNEL/ WEATHERED BEDROCK?)	Description unchanged by Amendment 1

FINE SOILS

Stiff red brown and grey green slightly sandy CLAY with some fine gravel and cobbles of siltstone (KEUPER MARL)	Stiff red brown and grey green slightly sandy gravelly CLAY with occasional cobbles. Gravel and cobbles are fine of siltstone (MERCIA MUDSTONE GROUP)	Stiff reddish brown and greyish green slightly sandy gravelly CLAY with low cobble content. Gravel and cobbles are fine of siltstone (MERCIA MUD-STONE GROUP)
Firm very thinly bedded brown very silty CLAY	Firm very thinly bedded brown CLAY/SILT	Firm very thinly bedded brown silty CLAY
Loose grey brown SILT	Uncompact grey brown SILT	Firm greyish brown SILT
Dense dark blue grey thickly laminated dark grey slightly fine sandy SILT. Colour changes to light grey quickly on drying (ESTUARINE DEPOSITS)	Compact dark blue grey thickly laminated dark grey slightly fine sandy SILT. Colour changes to light grey quickly on drying (ESTUARINE DEPOSITS)	Stiff dark bluish grey thickly laminated dark grey slightly fine sandy SILT. Colour changes to light grey quickly on drying (ESTUARINE DEPOSITS)
Soft thickly laminated grey CLAY with closely spaced thin laminae of grey fine sand with dustings of brown silt. Occasional pockets (up to 10 mm) of peat (ALLUVIUM)	Description unchanged by 1999 Code	Description unchanged by Amendment 1
Very stiff thinly bedded fissured and sheared dark grey mottled orange brown CLAY with frequent shell fragments. Fissures are generally subvertical very closely spaced smooth planar grey gleyed. Shears are subhorizontal (175/05 to 200/10) up to 500 mm persistence smooth undulating straight striated highly polished (Weathered LIAS CLAY)	Very stiff fissured and sheared thinly bedded dark grey mottled orange brown CLAY with frequent shell fragments. Fissures are generally subvertical very closely spaced smooth planar grey gleyed. Shears are subhorizontal (175/05 to 200/10) up to 500 mm persistence smooth undulating straight striated highly polished (Weathered LIAS CLAY)	Very stiff fissured and sheared thinly bedded dark grey mottled orangish brown CLAY with frequent shell fragments. Fissures are generally subvertical very closely spaced smooth planar grey gleyed. Shears are subhorizontal (175/05 to 200/10) up to 500 mm persistence smooth undulating straight striated highly polished (Weathered LIAS CLAY)

Table 2.9 *(Continued)*

BS 5930: 1981 description amended by Norbury *et al.* (1986)	BS 5930: 1999 description	EN ISO 14688 & 14689, BS 5930: 1999 Amendment 1 (2007)
Firm to stiff grey brown slightly fine sandy slightly gravelly CLAY with occasional lenses (5 by 15 to 15 by 50 mm) of yellow silty sand. Gravel is subangular to subrounded fine and medium of various lithologies (BOULDER CLAY)	Firm to stiff grey brown slightly sandy slightly gravelly CLAY with occasional lenses (5 by 15 to 15 by 50 mm) of yellow silty sand. Gravel is subangular to subrounded fine and medium of various lithologies. Sand is fine (GLACIAL TILL)	Firm to stiff greyish brown slightly fine sandy slightly gravelly CLAY with occasional lenses (5 by 15 to 15 by 50 mm) of yellow silty sand. Gravel is subangular to subrounded fine and medium of various lithologies (GLACIAL TILL)

MATERIALS AT THE SOIL/ROCK BOUNDARY

FINE	FINE	FINE
Very stiff brown grey and yellow brown mottled CLAY to very weak MUDSTONE very weak. Fissured and thickly laminated. Fissures are generally subvertical very closely spaced (Weathered LONDON CLAY)	Very stiff to very weak fissured thickly laminated brown grey and yellow brown mottled CLAY/MUDSTONE. Fissures are generally subvertical very closely spaced smooth planar with low polish (Weathered LONDON CLAY)	Very stiff to extremely weak fissured thickly laminated brownish grey and yellowish brown mottled CLAY/MUDSTONE. Fissures are generally subvertical very closely spaced smooth planar with low polish (Weathered LONDON CLAY)

COARSE	COARSE	COARSE
Light grey fine to medium SAND with occasional fine gravel size fragments of very weak light grey sandstone (BUNTER SANDSTONE)	Light grey fine to medium SAND with occasional fine gravel size fragments of very weak light grey sandstone (SHERWOOD SANDSTONE)	Light grey fine to medium SAND with occasional fine gravel size fragments of extremely weak light grey sandstone (D2 SHERWOOD SANDSTONE)

ROCKS

Brown medium grained thickly bedded moderately weathered dolomitic LIMESTONE strong (MAGNESIAN LIMESTONE).	Strong thickly bedded brown medium grained dolomitic LIMESTONE (MAGNESIAN LIMESTONE). Small but steady increase in strength with depth; colour mottled at top of stratum, effects due to weathering	Strong thickly bedded brown medium grained dolomitic LIMESTONE (CADEBY FORMATION). Small but steady increase in strength with depth; colour mottled at top of stratum, effects due to weathering
Light brown fine and medium grained thickly laminated to thinly bedded slightly weathered micaceous SANDSTONE very weak to weak (SHERWOOD SANDSTONE). Joints: dipping 45°, widely spaced, smooth planar stained	Very weak to weak thickly laminated to thinly bedded light brown fine and medium grained micaceous SANDSTONE (partially weathered SHERWOOD SANDSTONE). No weathering effects apart from staining. Joints: dipping 45°, widely spaced, smooth planar stained	Very weak to weak thickly laminated to thinly bedded light brown fine and medium grained micaceous SANDSTONE (partially weathered SHERWOOD SANDSTONE). No weathering effects apart from staining. Joints: dipping 45°, widely spaced, smooth planar stained

3 Systematic Description

Soils and rocks are made up of particles with a range of grain sizes and this range of sizes is categorised into a number of size fractions. These fractions are defined in terms of their size range, as discussed in Chapter 4.2. The fraction that controls the behaviour of the material is called the principal fraction. The characteristics of this principal fraction will be modified by any secondary constituent particles that are present.

In the description of soils and rocks, a very clear distinction needs to be made between the material and the mass features. A description covers the material characteristics and then follows with the mass features such as the discontinuities (Figure 3.1). The material is the intact soil or rock such as in a hand specimen without any discontinuities or weaknesses. The mass is the overall soil or rock including any discontinuities or other fabric present. A field exposure at some scale is necessary for a complete description of the mass features. Samples or cores from boreholes normally only allow a limited description of these characteristics.

The approach to the description of the material followed by the mass is, in principle, exactly the same in soils and rocks. However, as rock materials are so much stronger and stiffer than soils, the effect of the discontinuities is much more marked. The discontinuities and other fabric generally dominate the rock behaviour whereas in soils they affect or modify, but do not necessarily dominate, the overall behaviour. The description of the mass features such as the discontinuities is therefore critical in compiling the complete description. It is probably true to say that the description of the mass features could often be expected to take more time and effort than description of the material constituents.

It is pertinent in preparing to describe soils and rocks to consider the condition of the exposure being described and therefore how altered the exposure might be from the original or undisturbed in situ condition. It is useful to divide the exposure into three elements as follows:

- Nature of the grains: the particle size grading, shape and texture and plasticity. The nature of a soil or rock does not usually change during civil engineering works. The nature can be described on any quality of sample including disturbed (Quality Class 5) samples.
- State of the grains: the packing, water content, strength and stiffness. The state of the ground usually changes during civil engineering works. The description of the soil requires undisturbed samples or exposures.
- Structure of the mass: the fabric features of soils or rocks that are removed by reconstitution. Structure is often destroyed by large distortions and so can be observed only in the field or in an undisturbed (Quality Class 1) sample.

It is important not to over describe. In other words, to try to describe the structure of disturbed samples will at best be optimistic and at worst completely misleading. In the same vein, if the effort and expense to obtain Quality Class 1 samples has

(a)

(b)

Figure 3.1 Contrasting exposures: (a) rock material without any discontinuities, and (b) the more usual situation of a rock mass with very closely spaced discontinuities.

been made, the description should include all aspects of the state and structure that can be identified.

3.1 Standard word order

Field logs prepared by a logger in the field should utilise a defined terminology in a standard word order which are elements of the systematic process. This provides significant benefit for both the logger and the reader and also contributes greatly to the effectiveness of the communication as outlined below.

The standard terminology given throughout this book should be used in a consistent order of words. The purpose of a standardised word order is twofold. First, it enables the logger to work systematically and in an orderly fashion and it is to be hoped will therefore help them to remember to include all relevant aspects within the description. The preference for the logger to work in an orderly and consistent manner is discussed further below.

Second, the presentation of descriptions in a consistent word order is also important to enable the reader to access the information. The familiar word order

Table 3.1 Standard word order in description

Very coarse and coarse soil	Fine soil	Rock
(Relative density–where SPT N value available, see Chapter 5.1)	Consistency Discontinuities	Strength
Bedding	Bedding	Bedding
Colour	Colour	Colour
Secondary constituents in order of increasing dominance[1] (including particle dimensions, shape, strength, lithology, grading and composition for coarse and very coarse secondary fractions)	Secondary constituents in order of increasing dominance[1] (including dimensions, shape, strength, grading and composition for coarse and very coarse secondary fractions)	Secondary constituents as the matrix in clast supported materials or the clasts in matrix supported materials
Minor constituents	Minor constituents	Minor constituents
Particle dimensions, shape, strength, strength, grading and lithology of PRINCIPAL COARSE SOIL TYPE	PRINCIPAL FINE SOIL TYPE	ROCK NAME
Minor constituents	Minor constituents	Minor constituents
Other information	Other information	Other information
(GEOLOGICAL FORMATION including weathering classification)		
Additional sentences providing detail as necessary on principal or secondary constituents	Additional sentences providing detail as necessary on principal or secondary constituents	
Information on mass characteristics including weathering, fissuring or other fabric as detailed in Chapters 6 and 11	Information on mass characteristics including weathering, fissuring or other fabric as detailed in Chapters 6 and 11	Weathering description Discontinuities as detailed in Chapters 6 and 11 Fracture state

[1]Additional information on word order in the context of the description of secondary constituents is given in Chapter 8.1.

enables the key information and changes between strata to be more readily recognised. For example, the reader worried about the grading, or the strength, or the fissuring will know where to look for the information and to obtain the specific information that they are looking for.

No preferred word order is included in EN ISO 14688 or 14689 so national practice can continue as before. The standard word order that will continue to be used in UK practice is in accordance with BS 5930: 1999 as given in Table 3.1.

The standardisation of the word order allows introduction and use of a process flow chart as an aid to field practice as described in Chapter 4.3.

The word order used in rock description is the same as that used in soils. This was a change introduced in the 1999 code and was intended to make life easier in two ways. First, the logger only has to remember and work with a single word order and, second, the description of materials at the soil–rock boundary can switch smoothly between soil and rock terms as required. This is illustrated in Table 2.9 and discussed further in Chapter 4.10.

Examples of some typical soil and rock description are given in Table 2.9. These examples illustrate the use of a variety of approaches including the use of multiple sentences as outlined below to communicate information clearly.

3.2 The multiple sentence approach

In order to improve communication in descriptions, UK practice since 1999 has been to explicitly allow the division of a full description into more than one sentence. The intention of this approach is to make descriptions more readable by separating out detailed from less critical descriptive elements as a measure to help avoid unduly complicated descriptions. If a description is complicated because of a long string of adjectival descriptors, it is likely that it will include an element of ambiguity. This benefits no one and can be downright dangerous.

The first sentence of a multiple sentence description must name the principal soil type or name the rock and provide the quantification of the secondary constituents. The first sentence will also usually provide the overall information on the material. As many additional sentences as necessary are then used to provide any additional adjectival qualifiers on either the principal or the secondary constituents where required. This additional information can include description of the shape, grading, lithology or strength of particles (see Chapter 8.1 for more details on how to include this information). The additional sentences will also include details on the mass characteristics, including the weathering and full descriptive information on the character of discontinuities and/or fabric features.

Use of the multiple sentence approach is demonstrated below. First, an example is set out to illustrate the problem if multiple sentences are not used.

- 'Slightly medium to coarse shelly sandy subrounded fine to medium sandstone GRAVEL'

The problem with this description is that it is not clear whether it is the shells or the sand that is medium to coarse; it is easy to overlook the proportion of sand because the adjectives are separated and it is uncertain whether it is the sandstone or the gravel that is fine to medium. Whilst a bit of careful thought could resolve these uncertainties, there would always remain an element of doubt as to whether the logger and reader are communicating clearly and particularly whether we can be sure that this communication is without ambiguity.

It would therefore be better to recast this description as:

- 'Slightly sandy subrounded fine to medium sandstone GRAVEL. Sand is medium to coarse of shells.'

This version of the description is clear and unambiguous.

A further example of a straightforward description might be:

- 'Slightly medium to coarse sandy subrounded fine GRAVEL of medium grained sandstone'

In this case the description is clear and unambiguous so a second sentence is not needed. If a second sentence were to be preferred, this description could be rewritten as:

- 'Slightly sandy subrounded fine GRAVEL of medium grained sandstone. Sand is medium to coarse.'

A further example to illustrate varying approaches to how much information is included in the first sentence and therefore how many sentences are necessary could include the description:

- 'Slightly clayey subangular fine to medium sandstone gravelly generally medium to coarse shelly SAND with low content of subangular medium strong sandstone cobbles'

This description is correct but rather impenetrable. It may be that this could be more clearly written as:

- 'Slightly clayey gravelly generally medium to coarse shelly SAND with low cobble content. Gravel is subangular of fine to medium sandstone. Cobbles are subangular of medium strong sandstone.'

The simplification of the first sentence could be taken further such that the first sentence only includes the proportions of the various constituents, but this is not favoured:

- 'Slightly clayey gravelly SAND with low cobble content. Sand is generally medium to coarse of shells. Gravel is subangular of fine to medium sandstone. Cobbles are subangular of medium strong sandstone.'

The over simplification in the last version of this example is not encouraged. However, each company or project should decide on the approach to be used and stick to it. The key point is always that the description should be correct, clear and unambiguous.

In fine soils the approach is usually rather simpler to apply as follows, although the secondary constituent descriptors should still be included in the first sentence:

- 'Firm closely cracked slightly gravelly sandy CLAY. Gravel is rounded of quartzite. Desiccation cracks are rough planar unpolished.'

TIP ON BEST PRACTICE IN MULTIPLE SENTENCE DESCRIPTIONS

The multiple sentence approach is widely used, following the examples set out above, but without any consideration being given to the necessity or benefit of doing so. Whilst it is simpler to teach young loggers always to split the description of secondary adjectival descriptors into subsequent sentences, it is recommended that second and subsequent sentences should only be used where they are necessary to avoid ambiguity and to aid clarity. The general rule should be:

- Adjectives on the principal constituent are normally included in the first sentence along with the proportions of all secondary fractions.
- Adjectives on secondary constituents should be dealt with in additional sentences only where necessary for clarity.
- Adjectives on tertiary constituents are best included in the first sentence either before or after the principal soil type as dictated by common sense.

4 Description of Materials

4.1 Principal soil and rock types

The first step in any description is to identify the broad category into which the material will fall as this will determine the correct approach to the description. The various categories of soils and rocks are given in Table 4.1. At this stage of the process, the distinction between soil and rock is on the basis that rocks are of relatively high strength owing to being cemented or lithified whilst soils are relatively low strength uncemented particulate materials.

This initial decision needs to be made as the category to which a soil or rock belongs determines the approach used and some of the terminology varies slightly from one category to another.

The identification of which of the various categories is present in front of the logger and which they are going to describe can be arrived at using the flow chart presented in Figure 4.1.

Having identified the broad category, the description is then carried out in the sequence of the standard word order (see Chapter 3.2). However, although the rock or soil name is placed late in the description as presented on the field log, the identification of the rock or soil type actually has to be made as the first step.

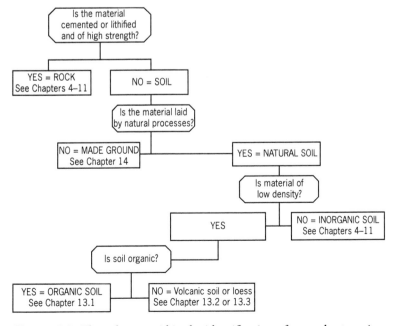

Figure 4.1 Flow chart to aid in the identification of ground categories.

Table 4.1 Categories of material

Sedimentary rock	Relatively strong material laid down by sedimentary processes from detrital material, chemical deposition or evaporation and cemented or compressed	Sedimentary rocks are described on the basis of their constituents and in terms of grain size
Igneous rock	Generally strong materials formed by igneous or volcanic processes	Igneous rocks are described on the basis of their mineralogy and in terms of their chemical composition
Metamorphic rock	Generally strong materials formed by alteration of other materials by heat or pressure or both	Metamorphic rocks are described on the basis of their constituents, grain size and fabric
Weak rock	Materials which are in the process of being lithified from soil to rock, or weathered from rock to soil	Materials transitional between soil and rock in terms of strength
Natural inorganic soil	Very coarse soil in which most particles, by weight, are larger than 63 mm	Boulders and cobbles which are described on the basis of particle size grading
Natural inorganic soil	Coarse soil in which most particles, by weight, are between 63 mm and 63 µm	Gravels and sands which are described on the basis of particle size grading
Natural inorganic soil	Fine soil in which most particles are less than 63 µm (and so cannot be seen with the naked eye)	Silts and clays which are described on the basis of plasticity as determined by hand tests
Natural inorganic soil	Soil is of low density and may be dark or other colours	Soils of volcanic (tuffs) or wind blown (loess) origin
Natural organic soil	Soil is of low density and of dark colour and may have organic odour	Peats which are described using material-specific terms
'Natural' man-made soil	Soil comprises any of the above but is placed by man	Made ground which can be described in accordance with rules for any of the above
Man-made soil	Soil is a manufactured material laid by man	Made ground which can be described in accordance with rules for any of the above classes as appropriate
Man-made soil	Soil is a manufactured material laid by man	Made ground which cannot be described in accordance with the rules for the above and material specific rules may need to be defined in each case

So, the broad identification of the rock name or the size fractions that make up a soil have to be identified and named in order that the appropriate terminology and features to be examined and described can be identified and followed. This chapter deals with this process of naming the soil or rock type. The description of the other features of soils and rocks is covered in Chapters 5 to 14.

4.2 Size fractions

All soils have a particular size fraction which dominates the behaviour of the soil and needs to be identified first. This can either be by this principal fraction being the dominant proportion of soil present in the case of very coarse or coarse soil, or by its plasticity in the case of fine soil. This dominant size fraction is termed the 'principal' soil type. This initial step of identifying the principal soil type is crucial as it sets the direction for all the following descriptive terms. Similarly in a rock, the principal constituents that control the behaviour of the rock material will be named first.

However, few soils are composed of only a single size fraction and other fractions are present which modify the behaviour of the principal fraction. These modifying fractions are the 'secondary' fractions. A soil can have more than one secondary fraction.

There may also be present minor or 'tertiary' components in a soil which are important because they assist in the identification of the origin and thus indirectly the likely behaviour of the soil. However, they do not, of their own accord, affect the behaviour of the soil.

Any constituent that does not fall into one of the above three categories does not need to be described, although such instances would be rare.

Examples of these three components of a soil are given in Table 4.2.

The description of secondary and tertiary fractions is given in more detail in Chapter 8.

In standard description terminology, soils come from one or more soil groups which are divided into size fractions. The groups are shown in Table 4.3.

The make up of a soil which comprises the coarse and fine size fractions can be plotted within the tetrahedral space shown in Figure 4.2. The corners of this tetrahedron are gravel, sand, silt and clay and therefore any combination of these four sizes will fall within the tetrahedron. Very coarse materials fall outside the tetrahedron and so these fractions cannot be plotted in this way. This separation reflects the impossibility of properly sampling and describing these fractions discussed in Section 4.4 below and is why they are described separately.

The behaviour of very coarse and coarse soils is controlled by gravity and interparticle forces. The particles can be picked up individually, examined by the naked eye, measured and described. The particles can be run through the fingers.

The boundary between fine and coarse soils at 63 micron is approximately the limit of visibility with the naked eye. The behaviour of fine soils is controlled by intermolecular forces which depend on the clay mineralogy present and the pore

Table 4.2 Examples of soil components

Example description	Principal fraction	Secondary fraction	Tertiary fraction
Stiff grey sandy CLAY with rare chalk gravel	CLAY – soil behaves as a fine soil	Sand modifies the plasticity of the fine soil	Chalk gravel identifies which till is present
Brown silty SAND with occasional brick fragments	SAND – soil behaves as a coarse soil	Silt reduces permeability of soil	Brick fragments show that stratum is made ground

Table 4.3 Particle size fractions

Soil type	Size fraction		Particle size
Very coarse soils	BOULDERS	Large boulders	>630 mm
		Boulders	630–200 mm
	COBBLES	Cobbles	200–63 mm
Coarse soils	GRAVEL	Coarse	63–20 mm
		Medium	20–6.3 mm
		Fine	6.3–2 mm
	SAND	Coarse	2–0.63 mm
		Medium	0.63–0.2 mm
		Fine	0.2–0.063 mm
Fine soils	SILT	Coarse	0.063–0.02 mm
		Medium	0.02–0.0063 mm
		Fine	0.0063–0.002 mm
	CLAY		<0.002 mm

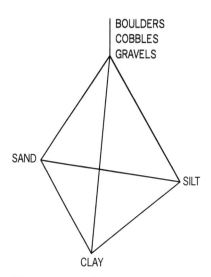

Figure 4.2 Soil tetrahedron.

water chemistry. Fine soil particles cannot be seen individually and distinction of and between fine particles is carried out by feel using a range of hand tests.

4.3 Description procedure using flow chart

A flow chart for the description of soils is presented in the Standards (Figure 1 in EN ISO 14688-1, Table 12 in BS 5930: 1999 Amendment 1). This flow chart is intended to lay out the steps that need to be followed in order to carry out the description of a sample. These steps form the logical and consistent process that is required in logging and goes together with the standard word order for descriptions. The process captured in the flow chart is described below.

It is easier to split this flow chart into two parts. The first part is given in Figure 4.1 and identifies the category of soil or rock that is present and that is going to

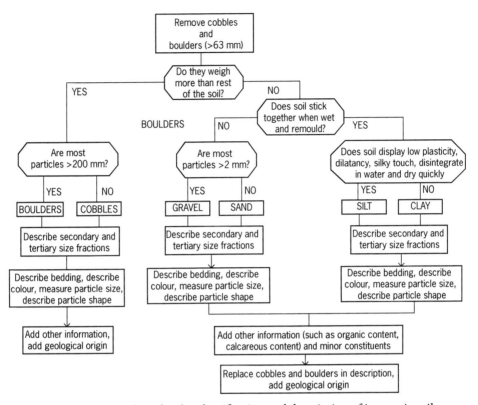

Figure 4.3 Flow chart for the identification and description of inorganic soils.

be described. The second part provides details of the process for logging the most common category which comprises the inorganic soils and is given in Figure 4.3.

The procedure to follow in accordance with Figure 4.3 is as follows:

1. Separate the very coarse fraction (all particles larger than 63 mm) from the rest of the soil. This is either a physical task or only a mental task if they are assessed and not removed.
2. Consider whether this fraction of very coarse particles weighs more than the rest of the sample.
3. If it does, the soil is very coarse. In this case it is either a BOULDER soil if most of the particles are larger than 200 mm or a COBBLE soil if most of the particles are between 63 and 200 mm in size.
4. The very coarse fraction is described. The material that is finer than 63 mm is now taken forward to step 5.
5. All soil particles smaller than 63 mm are taken for description as a coarse or fine soil with its constituents making up a 100% entity, whether this finer fraction is the predominant part of the soil or the secondary part of the soil.
6. For all the particles less than 63 mm, decide whether the remainder of the soil sticks together when wet. This may require addition of water to the sample as received. Note that the criterion of a soil 'sticking together' includes

the sample being remouldable. Fine sands at the right water content can cohere quite respectably, as shown by the possibility of sandcastle building on the beach, but cannot be remoulded.

7. If the sample does not 'stick together' when wet or does not remould, it is a coarse soil and is either a GRAVEL if most of the particles by weight are larger than 2 mm or a SAND if most of the particles are between 0.063 and 2 mm. This decision is of fundamental importance and is discussed further in Section 4.7.

8. If the sample does stick together when wet and remould, it is a fine soil. In this case the description is on the basis of plasticity not particle size grading. (This is in accordance with the fundamental precept of describing soils in accordance with their behaviour rather than in accordance with rules on grading or plasticity. This precept was originally laid down in BS CP 2001 (1957)). Simple hand tests for use in the field are required to determine whether the principal soil type is SILT or CLAY. These hand tests are described in Section 4.8.

9. Having decided on the principal soil type, the secondary fractions are described in terms of their proportion and effect on the particle size grading or plasticity as described in detail in Chapter 8.1.

10. Other aspects of the soil such as the consistency, bedding, discontinuities, colour and tertiary constituents are described.

11. Finally the 'very coarse fraction' and the 'coarse and fine fractions' are reunited and the descriptions linked to complete the description given below.

TIP ON OVERALL SOIL PROPORTIONS

- Care is required to ensure that the described combined coarse and fine constituent fractions do not exceed 100%, as discussed in Chapter 8.1.
- The coarse and fine fractions together are described as a 100% soil in their own right.
- When a soil includes very coarse particles these are separated out for the description as described above.
- The very coarse fraction is described as a 100% soil in its own right.
- To complete the description the very coarse and the finer fractions (coarse and fine together) are recombined.
- This subject is discussed in detail in Chapter 8.1.

The flow chart of Figure 4.3 described above clearly illustrates the process of description, but the format is not user friendly in terms of a crib for use in the field. Instead a tabulation of the process is commonly used as given in Tables 4.4 and 4.5. These tables are given in a format suitable for use in the field and present the process and word order in columns laid out in the correct order. The terms are also given with their definitions so there should be no excuse for incorrect terminology being used. The describer (or reader) should start at the left side and then proceed right across the table. On arriving at the right margin, the description should be complete and users should be comforted that the right terms have been used in the right order and with the right definitions applied.

In the description of soil, the initial four columns on Table 4.4 relate to the identification of the principal soil type. Once this important initial decision has been made, the user can proceed across the soil table along the series of rows commensurate with this identification. Thus in a coarse soil, the logger starts with relative density and concentrates the description on particle size grading and shape. In a fine soil the description starts with consistency and then includes more detail on discontinuities and bedding. The advantage of this layout is that in addition to the terms being given in the right sequence to suit the standard word order, definitions of the commonly used terms are provided. This sheet is therefore immediately useful and applicable in the field.

The principle of the rock table is exactly the same in that the logger proceeds from left to right describing both the material and then the mass features of the exposure with all terms being provided with their definitions.

These soil and rock description guidance sheets should be copied or printed out and laminated for field use. After all, logging is not a school exercise and it is not expected that any logger should try and memorise all the terms used in systematic description. Even if these terms are in daily use, it is more comforting to have the logger referring to a guidance sheet than not and it is a rare logger who will have the confidence to assert that all terms are correctly applied without an aide memoire to hand. Any sample description area or loggers clipboard that does not have these crib sheets fixed to the wall above the logging bench should give rise to a severe reprimand. Absence of these sheets is quite likely to result in erroneous descriptions.

TIP ON USE OF GUIDANCE TABLES

- Whenever you are going into the field to log it is strongly recommended that you take with you Tables 4.4 and/or Table 4.5 as necessary. It is handy if these are laminated to keep them serviceable for longer. These sheets are purposely A4 size so that they fit on a clipboard, but in a sample description area they can be enlarged and pasted to the wall. Anyone who is logging in the field without such guidance sheets should beware of thinking that they can remember the correct definitions of all the terms.
- If you find yourself in the field without these sheets do not worry as all is not lost. You should try to follow the standard word order, which should be well implanted in your brain. However, it may not be so easy to remember the definitions of all the terms. Do not over exert yourself in this matter, but rather record your observations in plain language with measurements in support. Thus record the penetration of your thumb as 15 mm rather than trying to remember if that should be soft or firm, give the bedding thickness as 125–210 mm and look up the correct term of medium bedded when you get back to the office.

4.4 Very coarse soils

Very coarse soils are those with particles greater than 63 mm across (boulders and cobbles) so the particles can be individually picked up and examined as hand specimens. An example of a boulder material is shown in Figure 4.4.

BACKGROUND ON LARGE BOULDERS

A recent change in European practice is the introduction of large boulders with a size in excess of 630 mm. This is reputed to be the limit of particles which can be picked up by a man (or at least a field geologist). As a particle of this size weighs of the order of 380 kg, there are not many geologists who would fit into this definition.

Table 4.4 Field Identification and Description of Soils

SOIL GROUP	PRINCIPAL SOIL TYPE	Particle Size (mm)	Visual Identification	Density/ Consistency Term	Field Test	Discontinuities
Very Coarse Soils	BOULDERS	large boulder / 630 / boulder / 200	Only seen complete in pits or exposures. Often difficult to recover whole from boreholes	None defined	Qualitative description of packing by inspection and ease of excavation.	Describe spacing of features such as fissures, shears, partings, isolated beds or laminae, desiccation cracks, rootlets etc.
Very Coarse Soils	COBBLES	cobble / 63				
Coarse Soils (over about 65% sand and gravel sizes)	GRAVEL	coarse / 20 / medium / 6.3 / fine / 2.0	Easily visible to naked eye; particle shape can be described; grading can be described.	Borehole with SPT N - Value		
				Very Loose	0–4	
				Loose	4–10	Fissured Breaks into blocks along unpolished discontinuities
				Medium Dense	10–30	
Coarse Soils	SAND	coarse / 0.63 / medium / 0.2 / fine / 0.063	Visible to naked eye; no cohesion when dry; grading can be described	Dense	30–50	Sheared Breaks into blocks along polished discontinuities
				Very Dense	>50	
Fine soils (over about 35% silt and clay sizes)	SILT	coarse / 0.02 / medium / 0.0063 / fine / 0.002	Only coarse silt visible with hand lens; exhibits little plasticity and marked dilatancy; slightly granular or silky to the touch; disintegrates in water; lumps dry quickly; possesses cohesion but can be powdered easily between fingers.	Very soft Finger easily pushed in up to 25 mm. Exudes between fingers		
				Soft Finger pushed in up to 10 mm. Moulded by light finger pressure		
				Firm Thumb makes impression easily. Cannot be moulded by fingers. Rolls to thread		
Fine soils	CLAY		Dry lumps can be broken but not powdered between the fingers; they also disintegrate under water but more slowly than silt; smooth to the touch; exhibits plasticity but no dilatancy; sticks to the fingers and dries slowly; shrinks appreciably on drying usually showing cracks.	Stiff Can be indented slightly by thumb. Crumbles in rolling thread Remoulds		
				Very stiff Can be indented by thumb nail. Cannot be moulded, crumbles		
				Hard Can be scratched by thumb nail		
				(or extremely weak)		

Scale of spacing of discontinuities

Term	Mean spacing, mm
Very widely	over 2000
Widely	2000 to 600
Medium	600 to 200
Closely	200 to 60
Very closely	60 to 20
Extremely closely	under 20

	Condition		Accumulated in situ
Organic Soils	Firm	Fibres compressed together	PEAT Predominantly plant remains, usually dark brown or black in colour, distinctive smell, low bulk density. Can include disseminated or discrete inorganic particles
	Spongy	Very compressible Open structure	Fibrous peat Plant remains recognisable and retain some strength. Water and no solids on squeezing
			Pseudo- fibrous peat Plant remains recognisable and strength lost. Turbid water and <50% solids on squeezing
			Amorphous peat No recognisable plant remains. Mushy consistency.
	Plastic	Can be moulded in hand. Smears fingers	Gyttja Paste and >50% solids on squeezing. Decomposed plant and animal remains. May contain inorganic particles
			Humus Remains of plants, organisms and excretions with inorganic particles

Bedding		Colour	Composite soil types (mixtures of basic soil types)		Minor constituent type	Particle Shape	PRINCIPAL SOIL TYPE	Minor Constituents
Describe thickness of beds in accordance with geological definition			For mixtures involving very coarse soils see BS 5930 Cl41.4.4.2			Very angular	BOULDERS	
						Angular		with rare
						Subangular		
						Subrounded	COBBLES	
						Rounded		with occasional
	Alternating layers of different types.	LIGHTNESS	**Term before principal soil type**	**Proportion secondary** *see Note 1*		Well rounded		with numerous/ frequent/ abundant
Inter-bedded or Inter-laminated	Prequalified by thickness term if in equal proportions.	Light Dark	slightly (sandy) *see Note 2*	<5 %		Cubic Flat Elongate	GRAVEL	
			(sandy) *see Note 2*	5–20% *see Note 3*				Terms can include:
	Otherwise thickness of and spacing between subordinate layers defined.	**CHROMA**	very (sandy) *see Note 2*	>20% *see Note 3*	Terms can include: glau-conitic micaceous shelly		SAND	shell fragments pockets of peat gypsum crystals flint gravel brick fragments rootlets plastic bags
		Pinkish Reddish Yellowish Orangish Brownish Greenish Bluish Greyish	SAND AND GRAVEL	about 50%	slightly (glauconitic)			
			Term before principal soil type	**Proportion secondary** *see Note 1*	(glauconitic)			
			slightly (sandy) *see Note 4*	<35 %	very (glauconitic)		SILT	
Scale of bedding thickness								
Term	Mean spacing, mm		(sandy) *see Note 4*	35–65 % *see Note 6*	Proportions defined on a site or material specific basis or subjectively			
Very thickly bedded	over 2000	**HUE**	very (sandy) *see Note 5*	>65 % *see Note 6*				
Thickly bedded	2000 to 600	Pink Red Yellow Orange Brown Green Blue White Grey Black			calcareous - clear but not sustained effervescence from HCl		CLAY	
Medium bedded	600 to 200			Terms used to reflect secondary fine constituents where this is important				Proportions defined on a site or material specific basis, or subjectively
Thinly bedded	200 to 60		Silty CLAY					
Very thinly bedded	60 to 20		Clayey SILT					
Thickly laminated	20 to 6				highly calcareous - strong and sustained effervescence from HCl			
Thinly laminated	under 6							

Transported mixtures		NOTES
Contain finely divided or discrete particles of organic matter, often with distinctive smell, may oxidise rapidly. Describe as for inorganic soils using terms above		1) % coarse or fine soil type assessed excluding cobbles and boulders
Term	Colour	2) Gravelly or sandy and/or silty or clayey
slightly organic	grey	3) Or described as fine soil depending on mass behaviour
		4) Gravelly and/or sandy
organic	dark grey	5) Gravelly or sandy
very organic	black	6) Or described as coarse soil depending on mass behaviour

Table 4.5 Identification and Description of Rocks

DIMEN-SION	ROCK MATERIAL					ROCK NAME			GENERAL	
	Strength	Structure and fabric	Colour	Texture	Grain size	SEDIMENTARY	IGNEOUS	METAMORPHIC	Minor Constituents	Formation Name
2000 mm	>250 MPa Extremely Strong — Rings on hammer blows. Only chipped with geological hammer	Use standard geological terms — Very thickly	LIGHTNESS	Use standard geological terms				Massive / Foliated	Describe using relative terms — rare	Name according to published geological maps and memoirs
630 mm	100–250 MPa Very Strong	Thickly	Light Dark			CONGLOMERATE	GRANITE		occasional	COAL MEASURES
200 mm	Requires many hammer blows to break specimen	Medium / Thinly laminated / Very narrow	CHROMA	for example / phaneritic	Coarse grained	BRECCIA / LIMESTONE		GNEISS	frequent	WEATHERED WICKERSLEY ROCK
63 mm	50–100 MPa Strong		Pinkish Reddish	ophitic		AGGLOMERATE	DIORITE	MIGMATITE	vugs	RESIDUAL MUDSTONE
20 mm	Rock broken by more than one hammer blow	Very thinly / Thickly laminated / Narrowly	Yellowish Brownish Greenish Bluish Greyish	porphyritic		DOLOMITE		MARBLE	shells / pyrite	
6.3 mm	25–50 MPa Medium Strong		HUE	crystalline		VOLCANIC BRECCIA	GABBRO		crystals	
2 mm	Cannot be peeled with knife, fractures with single blow of hammer	Thinly laminated / Very narrowly	Pink Red Yellow Brown Green Blue White Grey Black	amorphous	Medium grained — Coarse / Medium / Fine	SANDSTONE / HALITE / GREYWACKE / QUARTZITE / TUFF / ANHYDRITE	MICROGRANITE / DOLERITE / MICRODIORITE	SCHIST / PHYLLITE / QUARTZITE	organics / colours / odours	SHERWOOD SANDSTONE / LEWES CHALK
0.63 mm										
0.2 mm	5–25 MPa Weak									
0.063 mm	Can be peeled with difficulty. Point of hammer makes shallow indents									
0.02 mm	1–5 MPa Very Weak	Terms include bedded / laminated, foliated, banded and flow banded.			Fine grained	SILTSTONE / CHALK / GYPSUM / Fine grained TUFF	RHYOLITE / ANDESITE / BASALT	SLATE / HORNFELS		DISTINCTLY WEATHERED MERCIA MUDSTONE
0.006 mm	Crumbles under firm hammer blows. Can be peeled by knife									
0.002 mm										
	0.6–1.0 MPa Extremely weak — Gravel size lumps crush between finger and thumb. Indented by thumbnail				Very fine grained / Crypto-crystalline	MUDSTONE / Very fine grained TUFF / CHERT / FLINT	OBSIDIAN / VOLCANIC GLASS			

ROCK MASS										
DISCONTINUITIES										
Weathering	Orientation	Spacing	Persistence	Termination	Roughness	Wall strength	Aperture	Infilling	Seepage	No of sets
Approach 1 Mandatory description of all features associated with weathering	**Dip direction and dip** eg 245/70 Dip amount only in cores	Extremely widely > 6 m Very widely 2 to 6 m Widely 0.6 to 2 m	Very high > 20 m High 10 to 20 m Medium 3 to 10 m Low 1 to 3 m Very Low < 1 m		Large scale (m) Waviness Curvature Straightness	Use standard strength terms (Col 2) Support by using	Cannot normally be described in cores Extremely Wide > 1 000 mm Very Wide 100–1000 mm	Clean Surface staining (colour)		Cannot be described in detail in cores although division into sets with different dip amounts is often feasible and can be usefully carried out
Describe state and changes in Strength Fracture state Colour		Medium 200 to 600 mm Closely 60 to 200 mm Very closely 20 to 60 mm Extremely closely < 20 mm		x outside exposure r within rock d against discontinuity	**Medium scale** (cm) and small scale (mm) **Stepped** Rough Smooth Striated **Undulating** Rough Smooth Striated	Field strength tests Point load Schmidt hammer Other index tests	Wide 10–100 mm Moderately wide 2.5 to 10 mm	Soil infilling (describe as for soils, see over) Mineral coatings (eg calcite, chlorite, gypsum etc)	Moisture on rock surface Dripping water	
Presence or absence of weathering products			Discontinuous Continuous in cores	Cannot normally be described in cores	Smooth Striated **Planar** Rough Smooth Striated	Visual assessment	Open 0.5 to 2.5 mm Partly open 0.25 to 0.5 mm Tight 0.1 to 0.25 mm Very Tight < 0.1 mm	Other - specify	Water flow measured per unit time on an individual discontinuity or set of discontinuities Large > 5 l/sec	Record spacing and orientation of each set
Approaches 4 or 5 Classify only if useful and unambiguous		Take several readings Report minimum average and maximum		Record size of exposure	Measure amplitude and wavelength of feature	Take several readings Report minimum average and maximum		Record width, continuity and relevant characteristics of infill	Medium 0.5 - 5.0 l/sec Small 0.05 - 5.0 l/sec	

FRACTURE STATE

Solid Core — Solid core is taken as core with at least one full diameter (but not necessarily a full circumference) measured along the core axis or other scan line between two natural fractures.

TCR — Percentage ratio of core recovered (both solid and non-intact) to the total length of core run.

SCR — Percentage ratio of solid core recovered to the total length of core run.

RQD — Total length of solid core pieces each greater than 100 mm between natural fractures expressed as a percentage of total length of core run.

If — Average length of solid core pieces between natural fractures over core lengths of reasonably uniform characteristics, not core runs.

FI — Number of fractures per metre over core lengths of reasonably uniform characteristics.

Figure 4.4 Large (7 m) high boulder made up of boulders and cobbles cemented together.

Although the large particle size makes the description of these particles just like that of a hand specimen and therefore straightforward, it has to be borne in mind that most investigation methods in general use cannot recover such particles intact. This gives rise to two problems. First the particles are generally not recovered intact from the investigation hole and their presence in the ground needs to be inferred from indirect evidence, such as evidence of particle breakage, angularity and remarks by the driller. Second the sample presented for logging will probably not be large enough to be representative of the in situ soil, in particular once the particles include those much larger than about 100 mm (see below).

There is the added complication that the effect of even a small number of very coarse particles can be highly significant. For instance, the inclusion of a few boulders in a fine soil or flints in chalk can dominate the selection of, say, appropriate piling, tunnelling or excavation methods.

Very coarse soils are not common in Nature, which is fortunate as we are not really able to sample or test them sensibly. Reference to the size of samples required by Standards for testing in the laboratory, for example a particle size grading, puts the minimum quantity required for a representative sieving at 200 kg if a cobble is present or 1000 kg if a boulder is present in the soil. Although sieving of very coarse soils is not common, for which laboratories are grateful given these minima, strictly speaking a similar quantity should be available to the logger for the description to be representative. Clearly this is not possible from boreholes for two reasons. First it is not possible to recover very coarse particles from a standard borehole. Even in a 200 mm diameter borehole, the largest particle that can be recovered in the shell is only about 50–60 mm as shown in Figure 4.5. Second the total mass of soil

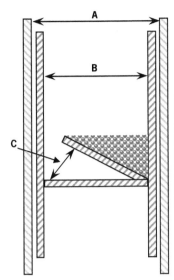

	Dimension	200 mm tools	150 mm tools
A	Inside diameter of casing	200 mm	150 mm
B	Inside diameter of shell	150 mm	100 mm
C	Aperture of flap opening once some soil particles are retained	60 mm	40 mm

Figure 4.5 Arrangement of boring tools when recovering coarse particles (courtesy Archway Engineering).

TIP FOR DESCRIBING VERY COARSE PARTICLES

- It is probable that few if any samples will be large enough to be representative if very coarse particles are present. If there are one or two cobbles or a single boulder present in the sample it is perfectly reasonable to report 'two cobbles' or 'one boulder of 280 mm'. This is more helpful than trying to apply the secondary quantifying adjectives given in Chapter 8.
- On the field log the information about the very coarse particles should be presented in summary form in the main description and the specific evidence for the presence of the cobbles and boulders should also be included in the detail indents. It is then possible for readers to assess the significance of these particles for themselves.

---

arising from 1 m of a 200 mm diameter borehole is only about 50 kg and so large enough samples cannot be achieved.

It is theoretically possible, of course, to obtain a large enough sample from a trial pit for the logger to be able to see a representative quantity. However, given the difficulty of excavation in such materials in terms of pit wall stability and probable water inflows, this is not usually practicable in reality.

It is usually best to use detail indents on field logs to present all information relating to the presence of very coarse particles, so that the reader can make their own assessment. The muckshift contractor, piling contractor or foundation designer will each have different views about the significance and impact of very coarse particles. It is therefore good practice to present the chiselling information, the driller's observations ('pushing boulder', 'breaking cobble'), the record of 'broken angular fragments' as important factual information. This will fit together with operator remarks on ease of excavation in the trial pit or in advancing the borehole. In addition, the main soil description can include a best estimate of the size and propor-

Figure 4.6 Particles on a 63 mm sieve. If these particles are turned and offered to the sieve opening, one particle is still retained and is a cobble while the other will pass though and is of gravel size.

tion of very coarse particles presented as outlined above and in accordance with the Standards.

A common problem in very coarse soils is being clear about which dimension is being measured. Very few loggers are able to answer this question, which is rather worrying. Answers offered range from the maximum to the minimum with all combinations of intermediate to average in between. In a laboratory particle size analysis the technician is required to offer the particle manually to the sieve aperture to see if it will fall through. The answer to the question is therefore that the size of any particle is the size of the square sieve aperture through which the particle would pass. The particle size is therefore not necessarily any of the sample dimensions. This is illustrated in Figure 4.6 where the lighter coloured particle is a cobble as it is too large to fit through the aperture, unlike the darker particle which does fit through and so is coarse gravel size.

4.5 Coarse soils

Physically, the description of coarse soils (gravels and sands) should be one of the easiest, as representative samples are recoverable from most types of investigation

---

BACKGROUND ON CHANGE IN GRAIN SIZE DEFINITIONS
- There have been many years of practice with the particle size definitions using the familiar 2–6–20–60 mm log cycles (see Table 4.3).
- This has recently been replaced by the series 2–6.3–20–63 mm to fall in line with international sieve sizes which are set at the 6.3 and 63 mm sizes.
- It has been proposed that it is not appropriate to interpolate between field descriptions and laboratory classification.
- In logging practice there is effectively no difference and so this does not change field practice.
- However, any reporting of size fractions on the basis of laboratory particle size analysis will need to use the correct size definitions.

hole in sufficient quantity for most tests and the particles can be seen by the naked eye and so are readily identified, measured and described.

There are two common problems in the description of coarse soils. The first is the difference between volume and weight percentages that needs to be accommodated. The human eye sees the proportions of sand and gravel by volume percentage. The heap of sand and the pieces of gravel occupying the same space are seen as the same proportion. However, the correct visual description is according to the weight percentage. This matter is discussed in Chapter 8.4.

The second problem in coarse soils is the correct application of the 2 mm boundary between sand and gravel. A sand/gravel mixture is shown in Figure 4.7 in which the separation of sand and gravel is clear, but not all soils are so straightforward. Most loggers have difficulty in identifying and naming particles above about 0.5 mm as sand let alone particles as big as 2 mm. If this error is not corrected, it is even more difficult to describe the proportions of the different size fractions correctly.

Questions that loggers of coarse and very coarse soils should be asking at the time of logging include the following:

- Have the largest particles been recovered or are there very coarse particles present in the ground?
- Is the 'sample' large enough to be representative, noting that whilst this only requires 17 kg if the largest particles are coarse gravel, this increases markedly if the largest particles are cobbles and boulders?
- Has the sample lost fines? Have the fines gone down the road in the dirty water returns arising from the drilling process or pumping, or were appropriate decanting procedures in operation?

In order to provide sufficient sample for a reliable description, the driller should be required to provide multiple samples. Similarly, an appropriate number of samples should be taken from trial pits to allow the anticipated testing to be carried out, even though the material has already been described at the pit before the samples are taken. It should be noted that:

- A bulk bag filled to capacity weighs about 20 kg.
- Take two well-filled bags to be sure of getting 25 kg or four for 50 kg.

Figure 4.7 Mixture of sand and gravel with a number of particles around the 2 mm size, which many loggers describe incorrectly.

- For actual quantities required for individual tests or combinations of tests refer to EN 1977-2 Annex L, where the requirements are very similar to those given in BS 1377: 1990, Part 1, Table 5.
- If a suite of earthworks testing (compaction, moisture condition value, strength) is being carried out, two or three bags will be required and more if the soil particles are degradable to enable fresh material to be used for each test.

4.6 Particle shape

The shape of particles of gravel, cobbles and boulders should be described using the terms illustrated in Figure 4.8.

There has been a tradition for many years of using four-fold terminology for the description of particle shape. This seems to have been perfectly adequate and

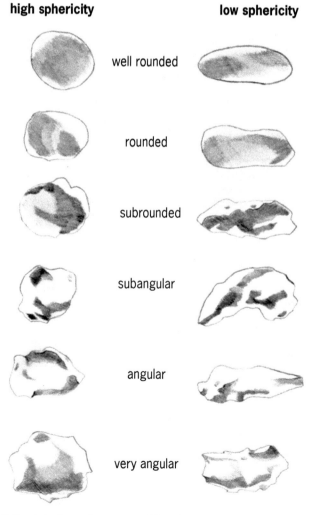

Figure 4.8 Particle shape descriptors (sketches prepared by Malcolm Williams).

has served well. EN ISO 14688-1 includes a classification with six terms, adding the terms 'well rounded' and 'very angular' in Figure 4.8. This approach is in line with the classification of particle shapes in sedimentological studies (for example Powers, 1953) but may be too refined for description for engineering purposes. It should be noted in this diagram that roundness refers to the sharpness and erosion of the corners and edges and that this is not the same as sphericity.

In addition to the terms above, terms such as 'flat', 'tabular' or 'elongated' can be given when the particles are far from being approximately equidimensional. These terms are only normally used in cases of extreme particle shape, or in aggregate studies, and are used in addition to the roundness or angularity terms.

TIP ON DESCRIPTION OF PARTICLE SHAPE
- In describing the particle shape, it is convenient to follow a convention that the descriptive terms are presented progressing from the angular towards the rounded. Thus 'angular to subrounded' for example.
- There is no particular reason for this other than same principles behind an overall word order.
- The shape descriptor given should be the most common shape or the general impression of shapes present. It is unhelpful always to include most or all of the range just because there are one or two particles falling towards the more extreme shape classes present.

The shape of sand grains can also be described. This is not usually of any benefit to the assessment of the engineering behaviour of the soil so that inclusion of this descriptor is not merited. There will be occasions, however, where the angularity of the sand grains is of importance and then there should be an explicit additional requirement to record this detail. Appropriate circumstances can include investigation for sand extraction or investigation of materials in a borrow pit on a site.

4.7 The coarse soil/fine soil boundary

The boundary between fine and coarse soils is possibly the most important boundary in soil description because it is the boundary between those soils which stick together and behave in a plastic manner when wet and those that behave as coarse non-plastic materials. This is the point at which we change from describing soils in accordance with the particle size grading curve, as determined in the laboratory (and assessed in the field by eye), and the plasticity, as identified by hand tests.

The boundary between coarse and fine soils is set at a particle size of 0.063 mm which is conveniently at about the limit of visibility to the naked eye. The description of fine soils (silts and clays) therefore has to rely on hand tests in order to ascertain the feel of the sample and certain behaviour and so the description is more operator dependent.

The key definition is the proportion of fine soil that that needs to be present to make the soil overall behave in a plastic manner, that is to stick together when wet and to remould. The argument as to the proportion of fine soil that relates to this boundary has persisted for years and remains elusive. This uncertainty is reflected in EN ISO 14688-1 where no definition is given; this is because at the time national representatives on the drafting committee were unable to agree a definition. The suggested criteria varied between about 35 and 50%.

The high impact of relatively small amounts of clay and to a lesser extent silt, in the overall behaviour of a soil is due to the higher specific surface of finer particles and the activity of clay minerals in this small size range. The specific surface is the ratio of surface area per unit weight. Surface forces are proportional to surface area (d^2) whilst self-weight forces are proportional to volume (d^3). The specific surface of clay minerals such as kaolinite or illite is several thousand times that of a quartz sand.

As the fundamental precept for soil description is to describe how the soils behave, the boundary is therefore not sensibly at 50%, but at some lower percentage of fines. In American practice, the issue is avoided and the boundary set at 50% based on sieving. Some European countries would prefer 40%, but UK practice has been to use about 35% fines. However, the actual proportion that controls the transition from a soil behaving as a fine soil to one behaving as a coarse soil is not a fixed figure.

This boundary figure depends on the overall shape of the particle size grading curve. A widely graded soil with a flat curve may need a proportion towards the top of this range because of the effect of the gravel particles which are relatively heavy as well as being inert in this context. In such a soil, a figure of 35–40% is normally reasonable. On the other hand a poorly graded soil with a steep grading curve may only need a much lower proportion of fines to control the behaviour because of the absence of coarse sand and gravel particles. For instance, a mixture of fine sand, silt and clay may require only about 20% fine particles to make it behave as a fine soil.

A general guide is that when a soil contains about 35% of fines that proportion may well be sufficient to control the soil behaviour. As long as the very coarse fraction is removed, this rule has been found to work reasonably well in practice for a large majority of soils. Consideration of theoretical packing arrangements of soil particles supports the 35% rule, although there is no agreement about whether the sand fraction should be included with the clasts or with the supporting matrix. This matter is discussed further in Chapter 8.3.

In BS 5930: 1981 the British Soil Classification System (BSCS) set the fine/coarse soil boundary at a fines content of 35%, but with no flexibility offered when it came to description. This lack of flexibility was perceived as inappropriate in soil description as there are many soils for which this is an inappropriate criterion (Norbury *et al.*, 1986). One of the reasons for disagreement arises from considering different aspects of the soil behaviour. As progressively more fines are added into a coarse soil, the fines initially fill the voids, affecting permeability but not strength. Once the proportion of fines reaches a certain limit, the coarse grains will no longer be in contact and the soil becomes matrix supported and the strength and compressibility are affected. Addition of further matrix separates the clasts further progressively reducing the shear strength. The point at which the fines fill the pores and seriously affect the permeability is different from the percentage at which the fines support and separate the coarse particles and thus control the shear strength and compressibility. In soil description, the differentiation is made on the basis of whether the 'soil sticks together when wet and remoulds'. This is probably an average condition between these various boundaries.

The assessment of this boundary in practice is illustrated in Figure 4.9 as discussed below:

- The grey soil on the right clearly sticks together, forms a mouldable ball and threads of about 3 mm diameter can be rolled. This is clearly a fine soil and is 'slightly sandy CLAY'.

Figure 4.9 Soils displaying a range of characteristics in response to the question 'Does the soil stick together when wet and remould?'

- The black soil on the left, on the other hand, can only be formed into a ball with difficulty and the ball is of poor quality with little coherence. The low fines content enables the soil to develop small suction pressures which are sufficient to hold the soil together. This is the characteristic of coarse soils that enables the building of sand castles. There is no chance at all of remoulding the soil by rolling a thread or manipulating the lump and so common

---

TIP ON THE NEED FOR FLEXIBILITY IN THE FINE–COARSE BOUNDARY

Considering the description of a number of example soils:

- 45% gravel, 30% sand, 15% silt, 10% clay. This material should probably be described as 'very sandy very clayey GRAVEL' as the 25% fines are not enough to control the behaviour, although this fraction would obviously influence the permeability and strength.
- 40% gravel, 25% sand, 20% silt, 15% clay. This material should probably be described as 'slightly sandy gravelly CLAY', the 35% fines being just about enough to support the clasts and control the material behaviour so that the soil will stick together when wet and remould.
- 80% sand which is mostly fine, 8% silt, 12% clay. The correct description is likely to be either 'very fine sandy SILT' or 'very fine sandy CLAY' depending on the clay mineralogy present. This soil is described as fine soil even though the fines fraction is only 20% as it will stick together when wet and remould.
- Considering a more complicated example, a soil comprising 12% cobbles, 50% gravel, 25% sand, 9% silt, 4% clay. The cobble content is excluded from the fractions used in the behavioural assessment and so the adjusted fractions are 57% gravel, 28% sand, 10% silt, 5% clay. The correct description could thus be 'very sandy clayey GRAVEL with medium cobble content'. This soil is still coarse in its behaviour because the fine soil matrix filling between the clasts is not sufficient to support the clasts and make the soil stick together when wet.

NOTE that the percentage term for fine soil in these examples is combined although only one term is used as it is not practical normally to achieve greater refinement. In other words the silt and clay terms remain mutually exclusive when present as secondary constituents in a coarse soil.

sense would conclude this soil is not fine. The soil is coarse and is in fact 'clayey gravelly SAND'.

• The red soil in the middle is more difficult. A good ball is readily formed and the soil sticks together when wet as required by the flow chart decision box (Figure 4.3). However, a thread cannot really be rolled. The attempts at threads shown in the photograph are over large at about 5–10 mm diameter and have been moulded rather than rolled. This soil is borderline as it sticks together when wet. In accordance with EN ISO 14688-1 this would be a fine soil. However, it does not remould so this soil should be described as a coarse soil. It is actually 'very clayey fine SAND'.

4.8 Fine soils

The allowable terminology for the description of fine soils, both individually and in mixtures, has not remained constant through the development of codification, as outlined in Chapter 2.7.1. Nevertheless the procedure and aim of the description of fine soils has not changed. The description of fine soils is based entirely on a set of hand tests, rather than on any measurement of the particle size grading. There are two main reasons for this.

First, the standard geotechnical laboratory test to determine particle size below the #200 sieve (75 μm which is the middle of fine sand) is sedimentation of some form. The calculation of the size is then based on Stoke's Law which relates to the settling velocity of spherical particles in a fluid. However, the clay and sometimes the silt fraction are not usually made up of spherical particles making the validity of this assumption tenuous. The grading curve obtained from a particle size analysis depends on the preparation procedure used and the assumptions made in the calculation of the results, although reasonably consistent results are obtained from the Standard test procedures.

Second, it should be remembered that these fine soils consist of very small particles which can include detrital particles, clay minerals and agglomerations of clay particles. Each of these can occur across the size ranges, so that we can get clay size rock flour particles and silt size aggregations of clay minerals. The clay and non-clay minerals present in the fine fraction behave in fundamentally different ways. Because of this variability of character and constituents, the description of fine soil should reflect how the fine soil fraction will behave, rather than being based on predetermined grading rules.

BACKGROUND ON MEASUREMENT OF PARTICLE SIZE IN FINE SOILS

Comparative studies of the particle size grading in clay tills using both standard sedimentation analyses and particle counting methods have been carried out on the author's projects. On the same sample the 'measured' clay content varied between 30 and 80%. The reason for the difference is not a fault of the tests, but was more likely to result from the measurements being made in different ways and the results being affected by the theory and statistical methods used to calculate the test results.

One of the main changes introduced in BS 5930: 1999 was a further attempt to clarify and simplify the terminology for the description of fine soils. Given that each of the Standards has introduced slight differences in this area (see Chapter 2.7.1) it is important that readers of old borehole logs remember this historical sequence of descriptive terminology otherwise they could become confused or, worse, misled.

The description of fine soils – the distinction between 'CLAY' and 'SILT' as a principal soil type – is made on the basis of a series of hand tests, namely:

- dilatancy
- toughness
- plasticity
- dry strength
- feel
- behaviour in water
- behaviour in air, and
- cohesion.

The procedure for carrying out these hand tests and thus the description of fine soils is described below. This series of tests is also described in ASTM D2487 but has been extended here to include a wider range of tests.

First, a representative sample of the material for examination is selected. The logger should then remove all particles larger than medium sand until a specimen equivalent to a small handful of material is achieved. This soil is moulded into a ball about 25 mm diameter until it has the consistency of putty; water should be added or the ball allowed to dry out as necessary in order to achieve the correct consistency.

Dilatancy

- Smooth the soil ball in the palm of one hand with the thumb, the blade of a knife or a small spatula.
- Shake horizontally, striking the side of the hand vigorously against the other hand several times. Alternatively the pat can be manipulated between the fingers of one or both hands (see Figure 4.10(a)).
- Note the speed with which water appears on the surface of the soil.
- Squeeze the sample by closing the hand or pinching the soil between the fingers as shown in Figure 4.10(b).
- Note the disappearance of the water film as none, slow, or rapid in accordance with the criteria in Table 4.6.

(a)   (b)

Figure 4.10 Carrying out the dilatancy hand test by (a) first jarring or shaking the pat, and (b) then squeezing it.

Table 4.6 for describing dilatancy

Description	Criteria
None	There is no visible change in the specimen.
Slow	Water appears slowly on the surface of the specimen during shaking and does not disappear or disappears slowly upon squeezing.
Rapid	Water appears quickly on the surface of the specimen during shaking and disappears quickly upon squeezing.

Toughness

- Roll the pat by hand on a smooth surface or between the palms into a thread about 3 mm in diameter as shown in Figure 4.11 (although the thread illustrated is larger than 3 mm). If the sample is too wet to roll easily, it should be spread into a thin layer and allowed to lose some water by drying. Fold the sample threads and reroll repeatedly until the thread crumbles at a diameter of about 3 mm when the soil is near the plastic limit as shown in Figure 4.12. Note the pressure required to roll the thread. Also, note the strength of the thread. After the thread crumbles, the pieces should be lumped together and kneaded until the lump crumbles. Note the toughness of the material during this kneading.
- Describe the toughness of the thread and lump as low, medium, or high in accordance with the criteria in Table 4.7.

Figure 4.11 Carrying out the toughness hand test and determining the stiffness of the rolled thread.

Table 4.7 Criteria for describing toughness

Description	Criteria
Low	Only slight pressure is required to roll the thread near the plastic limit. The thread and the lump are weak and soft.
Medium	Medium pressure is required to roll the thread to near the plastic limit. The thread and the lump have medium stiffness.
High	Considerable pressure is required to roll the thread to near the plastic limit. The thread and the lump have very high stiffness.

Figure 4.12 Carrying out the toughness and plasticity hand tests by rolling and rerolling the soil to form 3 mm threads.

Plasticity

- On the basis of observations made during the toughness test (Figure 4.12), describe the plasticity of the material in accordance with the criteria given in Table 4.8.

This test is the same as envisaged in EN ISO 14688-1 for the description of low plasticity or high plasticity fine soil. As this test is carried out in the distinction between CLAY and SILT, it is not necessary to use the plasticity terms as separate descriptors and this practice is discouraged. The application of the terms in Table 4.8 will satisfactorily report the outcome of this hand test and further terms are not necessary.

Table 4.8 Criteria for describing plasticity

Description	Criteria
Non plastic	A 3-mm thread cannot be rolled at any water content.
Low	The thread can barely be rolled and the lump cannot be formed when drier than the plastic limit.
Medium	The thread is easy to roll and not much time is required to reach the plastic limit. The thread cannot be rerolled after reaching the plastic limit. The lump crumbles when drier than the plastic limit.
High	It takes considerable time rolling and kneading to reach the plastic limit. The thread can be rerolled several times after reaching the plastic limit. The lump can be formed without crumbling when drier than the plastic limit.

Dry strength

- Make one or more test specimens comprising a ball of material about 12 mm in diameter. Allow the test specimens to dry in air or the sun. The ball can be dried in an oven as long as the temperature does not exceed 60°C.
- Test the strength of the dry balls or lumps by crushing as shown in Figure 4.13. Note the strength as none, low, medium, high, or very high in accordance with the criteria in Table 4.9.

Figure 4.13 Dry strength test showing low dry strength silt on the left and high dry strength clay on the right.

Table 4.9 Criteria for describing dry strength

Description	Criteria
None	The dry specimen crumbles into powder with mere pressure of handling.
Low	The dry specimen crumbles into powder with some finger pressure.
Medium	The dry specimen breaks into pieces or crumbles with considerable finger pressure.
High	The dry specimen cannot be broken with finger pressure. Specimen will break into pieces between thumb and a hard surface.
Very high	The dry specimen cannot be broken between the thumb and a hard surface.

- The presence of high-strength water-soluble cementing materials, such as calcium carbonate, may cause high dry strengths. The presence of calcium carbonate can be detected from the intensity of the reaction with dilute hydrochloric acid, see Chapter 8.2.

Feel

- The feel of the fine soils is different and the sensitivity of the fingertips in making this distinction should not be underestimated.
  - Clays feel smooth and readily take a polish when smeared with a blade or thumb. As the silt content increases, this propensity to polish decreases.
  - Silts feel silky and soft.
  - If there is an organic content, this imparts a more soapy feel to the soil.
- These tests also work well in making the differentiation between siltstone and mudstone/claystone.
- While holding the soil pat close to the ear, draw a steel knife or spatula blade through the soil and listen for the slightly gritty sound of the silt particles. Clay will give no sound of individual particles.
- It is still the case that undergraduates are being taught to use their teeth to determine the presence of silt. Whilst the silt grains do grate on the teeth and this is a valid test, this is not an approach to be recommended in these health and safety conscious days. It should be noted that many of the sites being investigated include contamination (natural or anthropogenic) to some degree.

Behaviour in air and water

- The stronger intermolecular forces present in clays which result from the smaller particles and the fact that they are clay minerals rather than small quartz particles, means that their behaviour in air and water differs. Clays stick together better, whether in air or water.
  - Prepare a ball of soil and place it in a bucket or tub of clean water. Silt will disintegrate within a few minutes, whereas clay will remain unchanged for hours, or even days as shown in Figure 4.14.
  - Smear some soil over a smooth surface (glass or plastic) or even better the back of the hand. Clay will remain wet, whereas silt will dry within a few minutes.
  - Once the soil is dry, it will not be possible to brush the clay off, because the smaller particles have settled into the pores of the skin, whereas the silt will brush off, leaving the hands clean. The drying process of clay is compared to that of silt in Figure 4.15.

Cohesion

The property of cohesion is demonstrated when compressing a ball or pat of soil.

- Compress the ball of soil between the fingers. Clay will deform in a plastic manner without rupture. Silt on the other hand will tend to crumble rather than deform as shown in Figure 4.16.

(a)

(b)

(c)

(d)

Figure 4.14 Behaviour of clay and silt in water. The silt in the left jar starts to fall apart reasonably quickly and the disintegration continues at a faster rate than the clay in the right hand jar. The first photograph (a) was taken 30 minutes after putting the soils in the water, (b) after about 1 hour, (c) after about 2 hours, and the last (d) about 4 hours later. This test can be accelerated with some mechanical agitation of the jars.

- While carrying out all these tests, it may have been necessary to add water to the soil, manipulating it in the hands and generally getting a feel for its behaviour. The final point to consider is how it has behaved during this time. The coarser the soil, the more water will have had to be added to keep the soil pat in prime condition for the tests.

Naming the soil as CLAY or as SILT is based on the results of these hand tests as shown in Table 4.10.

In the situation that all the hand tests point to the same conclusion of 'CLAY' or 'SILT', the decision is straightforward. When the various hand tests give conflicting results, one of the hybrid terms in Columns 3 and 4 of Table 4.10 should be used.

This scoring approach is not meant to be a hard and fast rule, more a suggestion as to how the principal soil naming can be carried out in practice. If there is doubt about the naming of the principal fine soil type or if it is critical to the investigation for any reason this should be taken as a reminder to carry out a comprehensive suite of laboratory index testing to aid thorough classification of the soil.

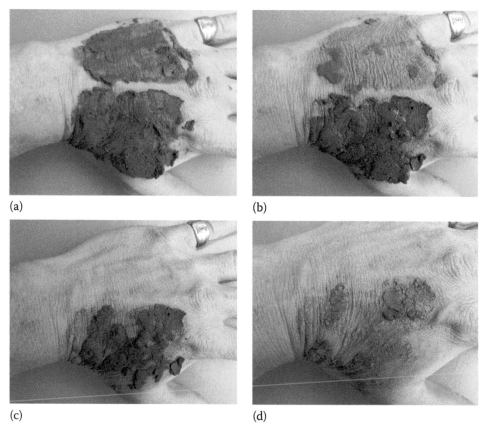

(a)   (b)

(c)   (d)

Figure 4.15 Behaviour of clay and silt in air. The grey clay dries much more slowly than the brown silt. Even when the clay has dried, it cannot be brushed from the hand whereas the silt has soon dried and can be brushed off. The drying in this instance took about 15 minutes, but would be faster in a warm low humidity environment.

Figure 4.16 Cohesion of fine soils. The silt on the left breaks and starts to crumble, whereas the clay on the right deforms without rupture.

Table 4.10 Identification of inorganic fine grained soils from hand tests

Soil description	CLAY	Silty CLAY	Clayey SILT	SILT
Dilatancy	None	None to slow	Slow	Slow to rapid
Toughness	High	Medium	Low to medium	Low or thread cannot be formed
Plasticity	High	Medium	Low	Non-plastic
Dry strength	High to very high	Medium to high	Low to medium	None to low
Feel	Smooth, sticky (when wet)	Smooth	Silky, sounds slightly gritty	Silky, sounds gritty
Behaviour in water	Disintegrates slowly if at all	Disintegrates slowly	Disintegrates in water	Disintegrates rapidly in water
Behaviour in air	Dries slowly with shrinkage	Dries slowly with shrinkage.	Dries quickly, brushes off	Dries quickly, brushes off
Cohesion	Deforms without rupture. Maintains shape and moisture during handling	Deforms without rupture. Maintains shape and moisture during handling	Moisture drains	Slumps, moisture drains

---

TIP ON NAMING BORDERLINE FINE SOILS

It is suggested that the fine soil be named in accordance with the following 'scores', being the results of the hand tests described above.

- When the 'score' is 6 or more to clay the soil should be described as CLAY.
- When the 'score' is 5–3 or 4–4 (clay–silt) the soil should be described as silty CLAY.
- When the 'score' is 3–5 (clay–silt) the soil should be described as clayey SILT.
- When the 'score' is 6 or more to silt the soil should be described as SILT.

There is mixed usage of the terms 'silty CLAY' and 'clayey SILT' in different parts of the world and historically (see Chapter 2.7.1). These terms are not considered available for use in BS 5930, implicitly in the 1981 code and explicitly in the 1999 code. This position has recently been changed by the phrasing of EN ISO 14688 which has reintroduced these terms into practice. The interpretation of this reintroduction, at least in the UK, is that as almost all fine soils are mixtures of clay and silt, the terms 'silty CLAY' or 'clayey SILT' should only be used where this is important, that is where the soil lies close to the borderline between the fine soils. This situation arises where the results of the hand tests described above are inconclusive. These terms therefore replace the hybrid term 'CLAY/SILT' introduced in BS 5930: 1999.

The recommendation here is for practice to go further and actively discourage the use of any additional adjectival terms such as 'very silty' or 'slightly clayey' in fine soil mixtures. Most practising loggers do not get the use of these two borderline terms anywhere near correct or consistent, let alone any more refined terms. It is considered better therefore not to use them. If the characteristics of the fine soil matter to the proposed works, it is better by far to carry out an adequate programme of index testing in order to characterise and classify the soil properly.

4.9 Classification of plasticity of fine soils

The distinction between CLAY and SILT in description is based on the hand tests described above. A check on this description and the formal classification comes from the results of index tests, known as Atterberg or consistency limits.

A convenient way to compare a variety of soils is to plot test results on a plasticity chart (Figure 4.17), in which an empirical boundary known as the 'A' line separates inorganic clays from silty and organic soils. Casagrande (1947) proposed the A line based on the information on soil properties available at that time. Soils of the same geological origin usually plot on the plasticity chart as straight lines parallel to the A line. The larger the plasticity index, the greater will be the volume change characteristics. 'Fat' or plastic clays plot above the line. Organic soils and silts generally plot below the A line. Clays containing a large portion of 'rock flour' (finely ground non-clay minerals) also tend to plot below the line.

However in his key note speech to the Guildford conference, Child (1986) pointed out that undue reliance on the A line was misguided. Many clay soils do indeed plot above the A line but some pure clays are found to plot below the A line. The situation becomes more complicated when either the clay size particles are of non-clay mineralogy or if the silt size particles are of clay minerals or clay aggregates. In either of these cases the empirical rules tend to break down. This matter has been considered by Polidoro (2003) who suggests that the plasticity chart needs to be considered afresh in the context of soil classification. This will not affect the description given in Section 4.8.

The terminology in EN ISO 14688-1 is based on a field assessment of plasticity followed by classification based on test results in EN ISO 14688-2. However, the same terms are used in the two soil standards with different meanings:

- Clauses 4.4 and 5.8 EN ISO 14688-1. The plasticity is described as being 'high' or 'low' on the basis of whether a 3 mm thread can be rolled. This test

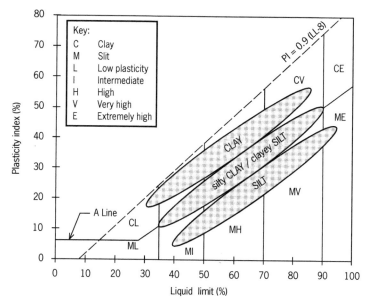

Figure 4.17 Plasticity chart showing plotted areas of clays and silts.

determines the plastic limit not the plasticity and the use of such plasticity descriptors would not normally be made. These tests have been described in Section 4.8 above.

- Clause 4.4 EN ISO 14688-2. The plasticity is classified into non-plastic, low, intermediate and high plasticity on the basis of the liquid and plastic limit, thus using the plasticity chart (Figure 4.17) based on the liquid limit. The use of such plasticity classification is perfectly acceptable, but this is not part of a field description. Such a plasticity classification is something that will only appear in a report text.

4.10 The soil/rock boundary

Geological processes mean that there are stiffer soils or weaker rocks which form the transition between rock and soil. Such materials can be present in the ground for three reasons:

- The soil material is in the process of lithification and has not yet reached the state of cementation of lithification to have the strength which would define it as a rock, to
- the material characteristics are such that a higher strength will not be reached because of high porosity, weak cementing or no cementing, to
- the rock is in the process of degrading to a soil.

An example of all three of these conditions is shown in Figure 4.18 which includes gravel which is indurated so as to have rock strength in situ but readily breaks down to the individual coarse soil particles. In addition, carbonate is being deposited locally through evaporation to give a cemented rock material.

Figure 4.18 Example of soil, apparent rock and cemented rock arising from a variety of post-depositional processes (Barzaman Formation, Oman Mountains; photograph courtesy of Steve Macklin).

A field description needs to enable the reader to identify clearly which of these processes is operating for the particular soil or rock material being described. This evidence and interpretation must be clearly presented to the reader as the geological implications and engineering properties of the ground in each of these three circumstances can be very different.

Some example specimens of soil and rock are presented in Figure 1.4. The coarse and very coarse soils in the centre and bottom left are clearly soil, although the cobble of limestone could be a hand specimen of a rock material. The core of fine soil in the top of the photograph is a marginal material being a very stiff CLAY/extremely weak MUDSTONE. The cores to the right of the photograph are clearly rocks.

Traditionally, the boundary between soil and rock was not defined, largely as a result of different approaches to description in the two classes of material. In BS 5930: 1999, the approaches to soil and rock were unified and a boundary defined on the basis of strength. If the material falls above this strength it will be described as a rock. If the material falls below this strength boundary it is described as a soil.

Soils have traditionally been described using the scale of shear strength (now called consistency) with the uppermost class being very stiff with a shear strength definition provided of >150 kPa. Conversely, the weakest rock description class was very weak with a compressive strength of <1.25 MPa (equivalent to 612 kPa shear strength).

Spink and Norbury (1993) suggested that the boundary be set at a shear strength of 300 kPa, which is of course equivalent to an unconfined compressive strength of 0.6 MPa. The reasons for the selection of this boundary value include that the boundary was sensibly in the middle of the previous range between 150 and 612 kPa. In addition, considering practical issues, the limit was found to be identifiable in the field as:

- the point at which a finger nail could no longer penetrate the surface;
- the strength at which fragments of material behave in a brittle rather than plastic manner in the hands;
- the degree to which a material will soften or disintegrate when exposed to water; and
- the rate at which this occurs.

This intermediate boundary was adopted in BS 5930: 1999 and was also later included in EN ISO 14688 and EN ISO 14689 although in these standards the recommendation comes as a suggestion in footnotes to tables where soils above, or rocks below, this strength can be described as the other, as appropriate.

The field identification of the soil/rock boundary is often not easy. It is common for materials at about the boundary to vary to either side of the boundary, whether their presence at this sort of strength is due to varying cementation or lithification or due to varying degrees of weathering. For this reason the word order in BS 5930: 1999 was made the same for soils as for rocks. It is thus possible to have a borderline material that is varying between soil and rock as 'very stiff CLAY/extremely weak MUDSTONE' (see Table 2.9).

It is relatively straightforward to describe this transition in fine grained materials as in the above example, but there is a shortage of appropriate terms for coarse soils. If standard penetration tests (SPTs) are being carried out, the boundary could be described as a 'very dense SAND/extremely weak SANDSTONE'. However, it may be that such test results are not available. A descriptive scheme to provide an assessment of the 'consistency' of sands is provided in Table 4.11. It will be interesting to see if any correlations become available with time of the relationship of this classification with SPT N values. As a preliminary suggestion it is considered that 'very dense' may equate to D1 and D2, whilst 'dense' may equate to D3 and D4.

It is suggested that the descriptive term from Table 4.11 be included with the geological formation name, in the same way as weathering classifiers, as there is no other logical position available in the word order.

The particular problem that affects all investigations in such materials is that this is the very range of materials at which it is most difficult to recover meaningful samples and even more difficult to preserve any samples recovered until safely in a testing machine. The description of very stiff soils and extremely weak rocks is therefore of critical importance in identifying the likely material behaviour, as test results may well not be available.

Table 4.11 Description of sand cores at the boundary with sandstone.

Term	Field definition	Penetration of thumb[1] (mm)
Extremely weak SANDSTONE	Thumb makes no impression. Core cannot be split with spatulas[2].	0
D1	Thumb makes little or no impression. Core can be split with two spatulas only with difficulty.	<1
D2	Thumb print makes clear impression. Core can be split but two spatulas required.	1–5
D3	Thumb makes deep impression. Core easily split with one spatula.	5–10
D4	Sand crumbles or disintegrates. Core splits readily with one spatula.	>10

[1] Penetration is measured by pressing the thumb firmly onto fresh surface of core or fragment.
[2] Spatulas refer to the normal blades or straight edges used for core splitting.

TIP ON CALLING A ROCK A ROCK

Many geologists carrying out logging of materials near the soil/rock boundary find it difficult to describe the material as rock when its strength exceeds the defined boundary. There is no obvious reason for this except that the formation name uses the word clay (e.g. London clay, Lias clay). Loggers should always be encouraged to make the decision and provide the correct description. After all, no one else has the material in their hand and is describing the strength of the fissure bound blocks. All subsequent strength determinations will be affected to some degree by the fissures or other fabric. Also, for the reasons cited, there may well be no test results available for any later reconsideration of this important and significant boundary.

4.11 Rock naming

In broad principle the naming of rock types follows the same principles as for soils. However, because of the greater induration or cementing, it is not usually practical to separate the grains into their constituent fractions. In addition, there is an entire geological nomenclature for naming of rocks arising from the geological sciences. This gives a different approach to naming based on the mineralogy or fabrics rather than the behavioural basis for naming soils. It remains appropriate however to follow this scientific practice wherever appropriate.

The level of detail that can be included in order to adopt the full range of petrographic terminology can be confusing for many geologists, let alone to the non-geologists who may also be using the field logs. It is thus appropriate to consider the level of sophistication that can and should sensibly be included in engineering geological logs. For most investigations a scheme of rock naming that is simplified from full geological practice is sufficient. This general scheme is presented below.

There are, however, rock-specific naming schemes used in engineering practice such as for coal, near-shore carbonate sediments which are widespread in the Middle East and volcanic rocks. These specific schemes are also described below illustrating more detailed material or formation specific schemes that can be established.

4.11.1 General naming of rocks

Tables of rock names that are commonly used in engineering investigations are provided below. The terms for sedimentary rocks are in Table 4.12 and the terms for igneous and metamorphic rocks in Table 4.13. These are provided in the Standards and Codes as an aid to rock identification. They are not meant to replace standard geological names, but a large majority of investigations can satisfactorily be reported using the more limited vocabulary indicated in these tables.

There will always be investigations where the use of a more detailed vocabulary would be sensible or even essential. In these cases more specialist knowledge than

Table 4.12 Aid to identification of sedimentary rock types

Grain size	Siliceous	Calcareous		Volcanic	Evaporites
>20 mm	Conglomerate		Calcirudite	Agglomerate	
6.3–20 mm				Volcanic	
2–6.3 mm	Breccia			Breccia	
0.63–2 mm	Sandstone	Limestone or Dolomite	Calcarenite	Tuff	Halite
0.2 mm–0.63 μm					
63–20 μm	Greywacke				Anhydrite
2–63 μm	Siltstone		Chalk	Fine tuff	Gypsum
<2 μm	Claystone Mudstone			Very fine tuff	
Crypto-crystalline	Flint Chert				Coal Lignite

> **TIP ON NAMING OF FINE GRAINED SILICEOUS ROCK**
> - The naming of fine grained siliceous rocks can be a problem. The logical name for these materials is claystone, to accord with the parallel terms used in naming soils.
> - In UK practice the term 'mudstone' needs to be used as the term 'claystone' has a long history and wide usage as the name for the nodular concretions which are commonly found in many fine grained siliceous rocks. This leaves a conflict with usage elsewhere as the term mudstone is reserved by some for rocks covering the clay (claystone) and silt (siltstone) size ranges, thereby avoiding the term siltstone.
> - This usage is not to be encouraged. Siltstones are clearly identifiable as a separate category of materials that behave differently from mudstone.

Table 4.13 Aid to identification of igneous and metamorphic rock types

	Igneous rock types			Metamorphic rock types	
Grain size	Light coloured (acidic)	Varying light to dark coloured	Dark coloured (basic)	Foliated	Massive
Coarse	Granite	Diorite	Gabbro	Gneiss	Marble
Medium	Micro-granite	Micro-diorite	Dolerite	Schist	Quartzite
Fine	Rhyolite	Andesite	Basalt	Slate	Hornfels
Crypto-crystalline	Obsidian	Volcanic glass			

is held by most practising engineering geologists may need to be invoked. More detailed knowledge could be available from national survey geologists, from the local university department or from geologists practising within the particular formation or basin.

This table is largely non-contentious. The bulk of the content is presented in Standards and has been in essentially this form for many years. The naming of volcanic rocks is discussed further in Section 4.11.4.

A simple set of terms for the description of igneous and metamorphic rocks is given in Table 4.13. The simplicity of this table would make the toes of many a petrologist curl, but most engineering geologists are not skilled in identifying rock materials from these sources. For this reason and also to enable reasonable communication with non-geological readers, the limited range of terms has much to commend it. It should not be beyond the ability of a logging geologist at least to put a rock name into the correct box in this table. Concern about whether the rock really is a diorite, or should more correctly be termed a grano-diorite or tonalite may not be critical to the engineering problem in hand. If it is considered that this naming is critical, appropriate expertise needs to be brought in and normal petrographic identification methods deployed.

It will be noted that the above table does not have a specific grain size scale unlike that for sedimentary rocks. All the standards except BS 5930 include a size scale. It was removed in the preparation of BS 5930: 1999 on the basis that a strict size scale flies in the face of normal petrographic practices. It is accepted that gabbro is a coarser grained acid rock than dolerite, but the boundaries are not at a particular crystal dimension of say 2 mm. The grain sizes in this table are thus relative and not specifically defined. This matter is discussed further in Section 4.12 in the context of grain size implications from rock names.

The logging of cores of rock or soil should also include diagnostic stratigraphic information such as marker beds and fossils. Markers can include identifiable and important changes of lithology or particular associations of deposit that are specific to that part of the succession. Marker features are often useful in correctly identifying the stratigraphic sequence and without this information the geological model is quite likely to be flawed or incomplete, as discussed further in Chapter 9.

4.11.2 Description of coal

Coal is a readily combustible rock containing more than 50% by weight of carbonaceous material formed from compaction and induration of variously altered plant remains similar to those in peat. After the application of considerable time, heat and burial pressure, these organic remains are metamorphosed from peat to lignite. Lignite is considered to be an 'immature' coal because it is still somewhat light in colour and it remains soil-like in character. The lithification process then continues with the lignite increasing in maturity, shown by its becoming darker and harder up to the point where it can be termed a sub-bituminous coal. After further burial and alteration, chemical and physical changes occur until the coal is classified as bituminous, recognisable as a dark hard coal. The hard coals are either termed 'bituminous' which is coal that ignites easily and burns long with a relatively long flame or the very hard and shiny 'anthracite'. This hardest coal represents the ultimate stage of maturity.

The full description of coal is a subject in its own right particularly when it is carried out as part of reserve studies. In this case, the calorific value and ash content need to be measured and the detailed studies that go with this usually include examination of petrographic thin sections. For the purposes of an engineering geological description, the constituents of coal can be considered to include the four material types shown in Table 4.14.

The coal will normally be described as black or brown coal and the inclusion of a coal type adds useful information. The description of the four constituent lithotypes can be carried out and the coal then classified as shown in Table 4.15 (Cameron, 1978).

Table 4.14 Coal constituent rock types

Lithotypes	Description
Vitrain	Very bright, glassy; bands or laminae a few mm thick; clean to the touch, tends to break into cuboids with conchoidal surfaces
Clarain	Lustre between that of vitrain and durain; striated texture due to alternation of bright and dull laminae
Durain	Grey to brownish black; rough surfaces with dull or greasy lustre; less fissured than vitrain
Fusain	Black or grey; silky lustre, fibrous structure, friable; only type that makes hands dirty.

Table 4.15 Classification of coal types

Coal type	Consituents
Bright	>75% vitrain plus bright clarain
Semi bright	50–75% vitrain plus bright clarain
Semi dull	25–50% vitrain plus bright clarain
Dull	<25% vitrain plus bright clarain

The level of detail to be provided and the nomenclature to be used in the description of coal should be agreed at the outset of any investigation.

In areas of coal workings, the presence or absence of coal seams is often the key issue. It is critically important to know where the coal seams should be in the succession and to make sure that their presence or absence is clearly and positively identified. Where the coal has been worked and assuming the workings have collapsed, there is likely to be a zone of broken ground above the level of the workings. Recording of these changed conditions will form the thrust of investigation in such areas.

4.11.3 Naming of carbonate sediments

The huge growth in construction activity in the oil rich states of the Middle East has required investigation and description of the particular materials present in that region. These are generally young near shore carbonate sediments laid down in a series of tropical and subtropical shelf seas and littoral shallows (Milliman *et al.*, 1974). Logging practitioners who tried to log using the rock naming schemes outlined in Table 4.12 produced a range of descriptions with various assorted discrepancies. The application of the normal approach in this geological environment did not provide sufficient refinement and accuracy for the requirements of engineering geological investigations. This matter was discussed by Fookes and Higginbottom (1975) who made some initial proposals. A more detailed scheme was subsequently proposed by Clark and Walker (1977) as outlined in Table 4.16. Although this scheme is not perfect it has become familiar and widely used in the region.

A recent review of the use of this scheme in the Barzaman Formation of the UAE (Macklin *et al.*, in press) has found that the scheme overall works reasonably well. However, care needs to be taken in identifying the nature of the cement which may be calcareous or dolomitic. It is therefore suggested that the term carbonate may be more appropriate than calcareous unless the proportion of the different minerals is known. The use of 10% acid on exposures is often satisfactory to assist this distinction (see Chapter 8.2). In addition care is needed in logging the brecciated fabrics which are common in that particular formation as a result of primary depositional and subsequent pedogenic processes which are continuing to this day. Two examples of these deposits are given in Figure 4.19 and Figure 4.20 which illustrate recent and continuing depositional facies over which are superimposed ongoing alteration, leaching and cementation.

4.11.4 Naming of volcaniclastic sediments

The descriptive scheme for volcanic sedimentary rocks given in Table 4.12 uses the terms breccia, agglomerate and tuff in order of decreasing grain size. These terms are descriptive but include implications about the genesis of the rocks and do not accord

Table 4.16 Scheme for naming of Middle Eastern carbonate soils (after Clarke and Walker, 1977)

Degree of induration	UCS (MPa)[1]	Additional descriptive terms based on origin of constituent particles				Total carbonate content (%)
		Not discernible		Bioclastic or oolite	Shell, coral, algal, pisolites	
		Clay size	Silt size	Sand size	Gravel size or coarser	
Non-indurated	<0.04 – >0.3	Carbonate mud	Carbonate silt	Carbonate sand	Carbonate gravel	>90
		Clayey carbonate mud	Siliceous carbonate silt	Siliceous carbonate sand	Mixed carbonate and non-carbonate gravel	50–90
		Calcareous clay	Calcareous silt	Calcareous sand		10–50
		Clay	Silt	Sand	Gravel	<10
Slightly indurated	0.3 – 12.5	Calcilutite	Calcisiltite	Calcarenite	Calcirudite	>90
		Clayey calcilutite	Siliceous calcisiltite	Siliceous calcarenite	Conglomeratic calcirudite	50–90
		Calcareous claystone	Calcareous siltstone	Calcareous sandstone	Calcareous conglomerate	10–50
		Claystone	Siltstone	Sandstone	Conglomerate	<10
Moderately indurated	12.5 – 100	Fine grained limestone		Detrital limestone	Conglomerate limestone	>90
		Fine grained argillaceous limestone	Fine grained siliceous limestone	Siliceous detrital limestone	Conglomeratic limestone	50–90
		Calcareous claystone	Calcareous siltstone	Calcareous sandstone	Calcareous conglomerate	10–50
		Claystone	Siltstone	Sandstone	Conglomerate	<10
Highly indurated	>70	Crystalline limestone or marble				>50
		Conventional metamorphic terminology to be used				<50

[1]UCS is unconfined compressive strength.

with recent practice. Brown (2007) outlined reasons for unease because the use of the above terms in this manner could result in erroneous assessment of the properties of such rocks. For example, the use of a term such as 'tuff' suggests a welded deposit of high strength. However many tuffs are actually weak rocks or are even unlithified and can readily devitrify and weather quickly. An alternative but still simple naming scheme that is correct in terms of current parlance proposed by Brown (2007) is given in Table 4.17. The criterion of a volcaniclastic rock containing more than 50% of grains being of volcanic rock as proposed by BS 5930 remains unchanged.

Figure 4.19 Excavation into wadi deposits comprising indurated very coarse soils in which carbonate cement is being deposited due to evaporation interbedded with fine overbank deposits. This could reasonably be described as 'calcareous CONGLOMERATE interbedded with clayey SAND' (photograph courtesy of Steve Macklin).

Figure 4.20 Metasomatically altered brown harzburgite cobble within caliche (carbonate cemented gravel) containing chert and green gabbro fragments; this could be described as 'congolmeratic LIMESTONE'; core diameter is 80 mm (photograph courtesy of Steve Macklin).

Table 4.17 Naming volcaniclastic rocks

Grain size (mm)	Current Standard terms	Terms proposed by Brown (2007)
>2	AGGLOMERATE if grains are rounded BRECCIA if grains are angular	Volcanic fragments in a finer matrix VOLCANICLASTIC CONGLOMERATE where fragments are rounded VOLCANICLASTIC BRECCIA where fragments are angular
0.06–2	TUFF – a cemented volcanic ash	VOLCANICLASTIC SANDSTONE
0.002–0.06	Fine grained TUFF	VOLCANICLASTIC SILTSTONE
<0.002	Very fine grained TUFF	VOLCANICLASTIC CLAYSTONE

This scheme seems to be a sensible clarification of some confused practice in earlier recommendations and so can be adopted as and when investigations are being carried out in such materials. The remainder of the material description in terms of strength, bedding and voids will complete the field record of these materials.

4.12 Grain size in rocks

The description of grain size in rocks is called for in all Standards but little or no guidance is provided on what is meant by this. Many rocks are composed of a single or very dominant single size or a small range of size of particles or crystals. The description of these rocks is straightforward following the procedures outlined above.

However, many rocks have a bimodal range of grain sizes, the most common examples being detrital grains in a matrix or larger crystals such as phenocrysts in a groundmass. In this case, both grain sizes should be described and the description should clearly identify their interrelationship, proportions and separate characteristics.

The effect of the proportions and thus the important grain size, in soils and in rocks should determine whether the rock material is clast (or crystal) supported or matrix supported. Whichever is dominant in this context will control the engineering behaviour of the rock material and should be the main rock name given as illustrated in the text box below.

EXAMPLE DESCRIPTIONS

Frequent quartz grains set in ferruginous cement	Sandy IRONSTONE
Sand with traces of quartz cement arising from pressure solution at grain contacts	Quartzitic SANDSTONE
Limestone with fine calcite cement and pelites	Shelly fine grained LIMESTONE
4–6 mm limestone crystals in calcite cement	Coarse grained LIMESTONE

The examples hint that there is an area of potential confusion regarding the grain size terms that are in common use. Standard grain size terms have to be used with caution as some rock names include the grain size connotations outlined in Table 4.18. This is not usually a significant problem and is unlikely to lead to major claims or failures in engineering works, but nevertheless care is required to ensure that the description is clear and not ambiguous in any way.

In the application of these grain size terms, some rock names which have no grain size connotation can be used without difficulty in accordance with these definitions. Thus, for example, LIMESTONE can be fine, medium or coarse grained and the above terms can be used as given without problem or confusion. A 'medium grained limestone' would therefore have crystals of between 0.006 and 2 mm.

However, other rock names have very definite grain size connotations, such as SANDSTONE. This is a medium grained rock as defined above, but it can itself be fine, medium or coarse grained, with size dimensions being the same as those defined for fine, medium or coarse sand. Thus, 'medium grained sandstone' has grains between 0.2 and 0.63 mm, not 0.06 and 2 mm as suggested in Table 4.18.

Table 4.18 Standard grain size descriptors

Term	Predominant grain size (mm)	Equivalent soil fraction
Very coarse grained[1]	Over 60	Cobbles and boulders
Coarse grained	2–60	Gravel
Medium grained	0.06–2	Sand
Fine grained	0.06–0.006	Silt (medium and coarse)
Very fine grained[1]	<0.006	Fine silt, clay

[1]See 'Background on grain size classes' below.

BACKGROUND ON GRAIN SIZE CLASSES

The terms 'very coarse grained' and 'very fine grained' are not commonly used in practice in rock description, having appeared in some of the early Working Party reports. These terms in fact map neatly across to the range of grain sizes used in soil description and can be clearly understood without confusion.

In accordance with common practice in rocks, if these terms are not used, the definitions of the coarse and fine terms respectively need to be adjusted to include the open ended size ranges.

The third case in this area is those rock names which do not have a particular dimension applied to their constituents at all. In the application of names from within geological science, GRANITE for example is a coarse grained rock but the boundary between granite and microgranite is not at 2 mm, as indicated in Table 4.13. This boundary is a qualitative difference which depends on the petrography in the context of the particular geological setting. As for sandstone, granites can still be fine, medium or coarse grained in a qualitative sense. Terms are to be used in this qualitative sense.

The outcome of this confusion is not actually too serious as long as common sense is applied in the description of rocks in the field and the expectations of the reader are managed sensibly. After all, in many investigations in igneous rocks, the field geologist describing the cores may well not be in a position to identify the rock types present correctly. This may require the help of petrographic examination and the results of this work will be in an annex to the report, not on the field logs.

5 Relative Density and Strength

The difference between description and classification as identified in EN ISO 14688 Parts 1 and 2 has been referred to in previous chapters. The field description covered within 14688-1 is for immediate use without test results, whereas classification covered by 14688-2 follows in the reporting process and uses test results from the field or laboratory. The logical conclusion of this clearly defined approach is that classification terms would not normally be expected to appear on a field log.

This approach is commended, not least as it provides simplicity and allows for more rapid completion of the field logs, as they are not waiting for the laboratory tests to be completed. There are also clear practical advantages in separating the field record from consideration of the results of tests.

However, there are a small number of classifiers based on test results which have traditionally appeared on field logs. There is a general feeling that to lose these classifiers from the field record would actually be a retrograde step. It is therefore suggested that certain classifiers can be used on field logs if there is benefit in doing so and provided always that no ambiguity arises.

The classifiers that have traditionally appeared on logs are relative density, strength and weathering grade. The first two of these are considered in the sections below and the application of weathering classifiers is discussed in Chapter 10.

5.1 Relative density in coarse soils

The standard penetration test (SPT) is an in situ test that is widely used in investigations in the determination of the relative density of a soil. The resultant N value can also be used to classify the relative density of coarse soils as part of the soil characterisation process.

The relative density term is a classifier, not a descriptor, being covered in EN ISO 14688-2 rather than in EN ISO 14688-1. The term cannot be assessed from examination of samples and so can only be added to a field log when SPTs have been carried out and using the test result.

It is suggested that custom and practice can be continued and that the relative density classifier can be included on borehole logs, bearing in mind the following caveats:

- Relative densities will not be given in the absence of an N value, so will not be included on trial pit logs.
- Any doubts about the validity of the N value blow counts for any reason, e.g. piping, must be noted on the borehole log.
- There should be no correction of either the N value or the relative density descriptor on the borehole log.

- If there are reasons for wanting to include any adjusted values on a field record, this should be additional information, with the uncorrected blow counts also being reported. The correction process followed must also be quoted, preferably on the log if at all possible.

The relative density of very coarse soils cannot be measured by SPT because the particles are larger than the drive shoe. The assessment of relative density in such soils is qualitative and based on inspection of packing and ease of excavation. Whether this will ever be possible to apply meaningfully is highly doubtful given the disturbing effects of creating the exposure.

Practice is to use the test result uncorrected in any way (N_{60}), although EN ISO 14688-2 provides no further guidance on the assessment of relative density. Classification terms are provided but are based on the result of laboratory maximum and minimum density tests. EN 1997-2 provides guidance in the informative Annex F with a classification of relative density based on N values corrected for hammer efficiency and overburden pressure, $N_{1(60)}$. These different classifications included in the EN Standards are summarised in Table 5.1.

It is generally considered that only straightforward factual information should be included on a field log. Thus the field log should only ever report the actual measured N value, with incremental blow counts. A relative density classifier can be included within the stratum description if so desired, but this should not be compulsory and should use the terms given in columns 1 and 2 of Table 5.1. The

---

TIP ON SPTs IN VERY LOOSE SOILS

It is often considered good practice to instruct the driller to double drive if the N value is less than 5, a problem which not uncommonly occurs in fine sands or silts. Double driving means carrying out a second test drive of 300 mm without withdrawing the SPT tools, giving an overall penetration of 750 mm. Whether this is a valued approach or not will depend on the particular project circumstances. On one hand this is a good means of ascertaining how much of the low blow count is due to boring disturbance and how much is actually genuine. Set against this is the use of a substantial length of borehole with no further test result or sample possible. If double driving found that the low count was the result of disturbance ahead of the borehole, the outcome is justified.

Table 5.1 Relative density descriptors for coarse soils

Practice on field logs		EN ISO 14688/2 Table 4	BS EN 1997-2 (Annex F)
SPT N	Term	Density index (%)	SPT $N_{1(60)}$
<4	Very loose	0–15	0–3
4–10	Loose	15–35	3–8
10–30	Medium dense	35–65	8–25
30–50	Dense	65–85	25–42
>50	Very dense	85–100	42–58

user of the field log is then free to make whatever adjustments they wish to the results or interpretation of them.

A complication arises, however, in that there is now a normative requirement for the energy ratio of dynamic probe (DP) (EN ISO 22476-2) and SPT (EN ISO 22476-3) hammers to be measured. If N values are to be used quantitatively, a calibration certificate is required to be available. The standards however provide no guidance on how to apply the measured energy ratio in design procedures.

A pro rata adjustment of the N value based on the energy ratio could be carried out as follows as suggested in EN ISO 22475-3.

$$N_{adjusted} = N_{measured} \times E_r/60$$

where E_r is the energy ratio.

However, this would require the data set to include a hammer identifier against each test result. This can be arranged in standard software, but is not currently available.

Further, is correction of the measured N values before use a sound proposal, when all the empirical correlations used on an almost daily basis were based on uncorrected values? The answer to this is probably in the negative. Early results from deployment in industry of wave analysers to calibrate hammers appears to confirm the results found by Clayton (1995) which is that even new hammers are only about 60–70% efficient.

Perhaps the only use of E_r will be for comparative assessment of test results. Such comparative exercises could help:

- in identifying those hammers that should not be in use as they are grossly inefficient,
- in comparing results when individual drillers produce different results; or
- in comparing results which vary through an investigation for reasons which could include changes arising from other factors, such as changes in temperature or wear of equipment.

Terms for classification of the relative density of soil exposed in trial pits are not included in the UK and European standards. There are simple tests for use in the field where the in situ state of the soil can be assessed. It is clear that any such assessment does not equate to the relative density. The use of such hand or field tests and classifiers may be suitable for some projects, such as in preliminary investigations or where rigs are not available to carry out boreholes and SPTs.

The first version of such an approach was provided in Anon (1972) as shown in Table 5.2.

Alternative definitions for assessing the density of coarse soils on in situ exposures, taken from USA practice, are given in Table 5.3.

The results of these hand tests do not replace the results of standard tests carried out in boreholes. In accordance with the precept outlined in EN ISO 14688, if the results are to be used merely qualitatively, the above hand tests provide useful information. However, if the assessment is to be used quantitatively in any way, such as in the assessment of bearing capacity or settlement, the relative density will need to be measured by carrying out SPTs at some stage in the investigation.

Table 5.2 Field density descriptors

Term	Definition	Comment
Indurated	Broken only with sharp pick blow, even when soaked. Makes hammer ring.	Should probably be described as medium strong to weak rock.
Strongly cemented	Cannot be abraded with thumb or broken with hands.	Should probably be described as weak to very weak rock.
Weakly cemented	Pick removes soil in lumps, which can be abraded with thumb and broken with hands.	May be described as extremely weak rock.
Compact	Requires pick for excavation; 50 mm wooden peg hard to drive more than 50–100 mm.	
Loose	Can be excavated with spade; 50 mm wooden peg easily driven.	

Table 5.3 Descriptive terms for reporting density of coarse soils in situ after McCook (1996) and BRE (1993)

Descriptive term	Ease of excavation	Field test using reinforcing rods	Field tests using a 1.2 m long, 38 mm diameter auger
Very loose	Very easy to excavate with a spade.	No field test definition provided.	The auger can be forced several inches into the soil, without turning, under the body-weight of the technician.
Loose	Fairly easy to excavate with a spade or to penetrate with a crowbar.	Easily penetrated with a 13 mm diameter reinforcing rod pushed by hand.	The auger can be turned into the soil for its full length without difficulty. It can be chugged up and down after penetrating about 300 mm, so that it can be pushed down 25 mm (1 inch) into the soil.
Medium dense	Difficult to excavate with a spade or to penetrate with a crowbar.	Easily penetrated with a 13 mm diameter reinforcing rod driven with a 2.3 kg hammer.	The auger cannot be advanced beyond 750 mm without great difficulty. Considerable effort by chugging required to advance further.
Dense	Very difficult to penetrate with a crowbar. Requires a pick for excavation.	Penetrated 300 mm with a 13 mm diameter reinforcing rod driven with a 2.3 kg hammer.	The auger turns until tight at 300 mm; cannot be advanced further.
Very dense	Difficult to excavate with a pick.	Penetrated only a few centimetres with a 13 mm diameter reinforcing rod driven with a 2.3 kg hammer.	The auger can be turned into the soil only to about the length of its spiral section.

An alternative set of terms has been given in Chapter 4.5 primarily in the context of identifying the soil/rock boundary, but the terms D1 to D4 could also be considered for use in describing the condition of the ground. The reasons for a coarse soil material holding together include the presence of cement, grains

---

TIP ON USE OF NON STANDARD FIELD TESTS

- If tests such as those given in Table 5.2 and Table 5.3 are used the results should be included together with the test method on the field log. The application of a relative density term should be provided as extra information, not as the first item in the first sentence of the description.
- For example, the material exposed in a trial pit could be described as:
  - 'grey SAND. Assessed as loose using 38 mm auger' or
  - 'brown sandy GRAVEL. Dense based on reinforcing rods penetrating up to 200 mm'.

---

being locked, and bonding. The cause and influence of these different processes is a matter of considerable ongoing debate in the literature, but that uncertainty should not affect the logger describing the in situ condition of these materials in the field.

5.2 Consistency of fine soils

Historically the description of 'strength' has been used in field logging. This traditional description of soil 'strength' was an imprecise approach as it mixed description and classification together. This imprecision has allowed specialists and non-specialists to abuse the so called strength descriptor on field logs by taking, say, an allowable bearing pressure direct from the 'strength' given on a field log. A change introduced by EN ISO 14688 amends this procedure for the better. There is now an implicit requirement to separate the reporting of 'strength' into the field recording of consistency and the subsequent classification of strength on the basis of test results.

Consistency is a field descriptor and is defined in EN ISO 14688-1. The soil may also be classified into strength categories as defined in EN ISO 14688-2. The strength classifier is not a descriptor and so its inclusion on a field log is not a mandatory requirement.

The traditional approach prior to the EN ISO has been to use the consistency and strength terms given Table 5.4 and Table 5.5 together. The new approach given in the EN ISO is to describe consistency in the field using the terms and definitions given in Table 5.4.

The procedure in these hand tests is illustrated in Figure 5.1.

Description of the consistency of fine soils should only be carried out on samples that are undisturbed and considered by the logger to be representative of the in situ ground condition. The description of consistency of lower quality class samples can be made, but only for information at the logging stage. The only use of such observations is the likelihood that the soil in situ is at least as stiff as that observation.

Soils with a consistency greater than very stiff would normally be expected to be more appropriately described as extremely weak rock, unless they are transported soils. This soil–rock boundary is identified by Spink and Norbury (1993) where the distinction between the consistency of very stiff clay and extremely weak mudstone is taken as the point above which it is no longer possible to indent the clay with the thumbnail (the thumbnail can scratch the surface) and the material breaks in a brittle rather than plastic fashion.

The description of the consistency of silt should also be described using the same terms as given in Table 5.4. However, there are two problems. The first is that

Table 5.4 Terms for describing consistency

Term	Field definition
Very soft	Finger pushed in up to 25 mm, exudes between fingers.
Soft	Finger pushed in up to 10 mm, moulds by light finger pressure.
Firm	Thumb makes impression easily, cannot be moulded by fingers, rolls to thread.
Stiff	Can be indented slightly by thumb, crumbles, breaks, remoulds to lump.
Very stiff	Can be indented by thumbnail, crumbles, does not remould.
Hard (transported soils only)	Can be scratched by thumbnail.

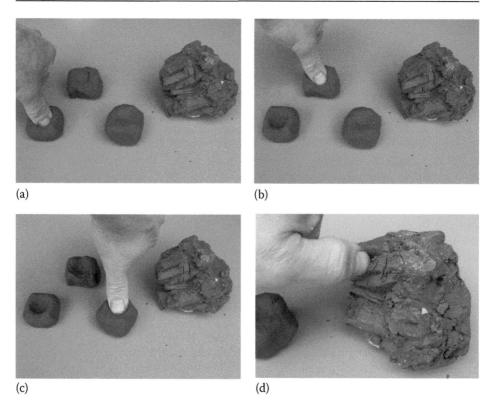

(a) (b)

(c) (d)

Figure 5.1 Hand tests to determine consistency of fine soil: (a) 'Soft CLAY' where the finger can be pushed in easily to 10 mm which is about the length of the thumbnail; (b) 'firm CLAY' as shown by finger easily making a significant impression; (c) 'stiff CLAY' where the finger has made only a shallow impression; (d) 'very stiff CLAY' with the thumb making no impression, but the thumbnail making indentations which are visible to the right of the thumb.

the hand tests defined in Table 5.4 are not readily applicable to silts, particularly if the silt is medium or coarse and therefore tending to behave as a coarser soil. The second is that obtaining a representative sample of silt is very difficult and in practice the consistency is usually changed by the sampling. Driving a sample tube

TIP ON DESCRIBING CONSISTENCY OF FINE SOILS
- Appropriate samples for description of consistency can include rotary cores, tube samples and material from the shoes of tube samples, in situ material seen in trial pits and freshly excavated material in large pieces or clods in the excavator bucket.
- It is possible to obtain reliable descriptions of consistency from samples which are recovered from the shoe of a tube or core sampler.
- Reliable descriptions of consistency are unlikely to be obtained from disturbed samples such as bag, jar or grab samples or from SPT shoe samples.
- At the description stage, the consistency of these disturbed samples can be noted for completeness, but it is suggested that this descriptor be given (in brackets) to make distinct the poor quality of information provided.
- Consistency descriptors from disturbed samples would not normally be transferred to the field log.

TIP ON DESCRIPTION OF HARD SOILS
The use of the consistency term 'hard' should be reserved for transported soils which have not been stiffer such as glacial tills/boulder clays. If the soil is not transported, as is the case in the solid strata (in geological map terms), the clay will be at a strength that would mean its strength is high enough to behave as a rock and the equivalent term that should be used instead is 'extremely weak'.

TIP ON DESCRIPTION OF CONSISTENCY OF SILT
- As a result of the difficulty in sampling silts without significantly altering their state, the terms obtained using the tests in Table 5.4 should be used with caution.
- Unless the logger is confident that the terms provided are meaningful and representative of the in situ ground condition, the consistency terms for silt should be omitted from field logs.

can either loosen or densify the silt depending on grading, water content and tube geometry. It is likely that a soft silt will be densified by sampling and a stiff silt loosened by sampling unless special sampling measures appropriate to silt soils have been adopted.

5.3 Strength: shear or unconfined

Having described the consistency of a fine soil in the field, the next non-mandatory step is to classify the strength solely on the basis of test results. These results may be obtained in the field (using a vane for example) or in the laboratory by strength index or triaxial tests. The classification of strength is carried out using the terms given in Table 5.5.

Given the practise in which the boundary between soil and rock is set at a shear strength of 300 kPa, the classification term 'extremely high strength' should only be used for transported soils such as glacial deposits. If the soil is not transported and is, for example, a mudstone or sandstone, the rock strength term 'extremely weak' should be used instead.

Table 5.5 Terms for classification of shear strength

Term	Definition (kPa)
Extremely low	<10
Very low	10–20
Low	20–40
Medium	40–75
High	75–150
Very high	150–300
Extremely high	>300

TIP ON MATERIAL STRENGTH
- The inclusion of strength terms on field logs was proposed by Baldwin *et al.* (2007). The approach suggested was to include the strength classifier after the discontinuity descriptor. Thus for example 'stiff closely fissured medium strength slightly brown sandy CLAY'.
- It should be noted that the field consistency in this example is probably one class higher than the measured strength.
- This apparent discrepancy reflects the difference between the intact material strength between the fissures (consistency) and the laboratory test which includes some defects such as fissures in the tested volume (strength).

EN ISO 14688-2 does not require the strength classifier to appear on field logs. However, there is no particular reason why this is not possible as an enhancement of this approach, adopting a similar logic to that used to support the inclusion of relative density classifiers on field logs.

However, it is noted that although this suggestion is apparently reasonable, its application is not straightforward in practice. The measurement of strength is on individual samples and, for a number of reasons, the result of that test may be considered unrepresentative of the stratum as a whole. Reasons for this can include:

- sample disturbance causing a decrease in strength;
- sample aging or drying causing an increase in strength;
- differences arising from scale and the effects of fabric;
- the sample not being representative of the whole stratum; or
- changes in the strength of the stratum with depth.

In addition, there may not be test results available through the full depth of individual strata or on all strata and so the approach that would be found on a finished field log would be inconsistent as to whether and how the terms are provided.

Many investigations at shallow depth, say for spread foundations, are in the depth range where there are rapid changes in strength with depth. The rate of change may not be captured by a typical frequency of testing so that a strength term based on too few test results could very easily be misleading. It is not recommended that this possibility be adopted.

Early investigations of attempts to include strength classifiers on field logs after Baldwin *et al.* (2007) seem to indicate that the classifier is left off more often than it is appropriate for it to be included, as a direct outcome of the problems identified above. Inclusion of the test results and the strength classifier in the text of a report where the appropriate parameters are being derived would be preferable, so that the validity or otherwise of the test results and the reasons for any problems with the results and the consequent classification can be discussed.

One of the most common areas of discrepancy between field descriptions based on hand tests in comparison with test results is that arising from different sample sizes. The effect on measured strength of larger samples, which include more defects such as bedding or fissures, is well known. Description in the field is carried out on hand specimens so that the material strength is determined, that is the strength in between the fissures or joints. This strength is likely to be higher than the strength measured in a 100 mm diameter sample or a larger volume encompassed by an in situ test such as a plate loading test.

A correlation between undrained strength and SPT N value was proposed by Stroud (1989). The author proposed a relationship:

$$c_u = f_1 \times N$$

where f_1 is a factor dependent on plasticity.

The value of f_1 is commonly in the range 4.5–5.5. On this basis a correlation or check of strength and N value can be made in accordance with Table 5.6.

Strengths assessed in this manner should not appear on field logs. They are classifiers and suffer from the problems identified above. They should therefore act as no more than a useful cross check against other test results.

The approach to the description of strength in rock is different from soil in that it continues with the earlier system of combining field descriptive terms and strength classifiers based on test results. The descriptive and classification terms given in EN ISO 14689-1 are shown in Table 5.7. As there is only a single set of terms there can be no confusion about what terms appear on a field log. Fuller field definitions of these terms are given in Table 5.9.

When using Table 5.7 it is always good practice to check field strength descriptors against available test results. However, there is the same problem here as in soils because of the difference in strength between the material and the mass. The field description of strength is that of the material in between the joints. Measure-

Table 5.6 Strength classifiers from N value

Strength	N value (blows/300 mm)
Extremely low	N value unlikely to be meaningful or reliable
Very low	<4
Low	4–8
Medium	8–15
High	15–30
Very high	>30
Extremely high	No value given

Table 5.7 Rock strength description and classification terms

Term	Definition for field use	Unconfined compressive strength (MPa)
Extremely weak (or hard soil)	Indented by thumbnail	0.6–1.0
Very weak	Crumbles under firm blows with geological hammer, can be peeled by a pocket knife.	1–5
Weak	Can be peeled by a pocket knife with difficulty, shallow indentations made by firm blow with point of geological hammer.	5–25
Medium strong	Cannot be scraped with pocket knife, can be fractured with single firm blow of geological hammer.	25–50
Strong[1]	Requires more than one blow of geological hammer to fracture.	50–100
Very strong[1]	Requires many blows of geological hammer to fracture.	100–250
Extremely strong[1]	Can only be chipped with geological hammer.	>250

[1]Wear eye protection when using hammer.

TIP ON FIELD ASSESSMENT OF ROCK STRENGTH
- The hammer referred to is a standard 0.9–1.0 kg geological hammer, not a sledge hammer.
- The specimen should be resting on a solid surface unless stated otherwise.
- It is not necessary always to break the rock specimens. It is much safer, and a lot easier, to tap the rock firmly with the hammer or a knife blade and describe on the basis of the sound that the hammer makes.
- As the strength increases, the sound goes from dull to bright, increases in pitch and changes in the tone from whop to thwack to ploink to plink to dink. No reliable confirmation of these terms is available at present.

ment of strength by whatever test method will use larger specimens, which will usually include some defects so the mass strength measured will be lower than the material strength.

Measurement of the strength in rocks can use a variety of tests, most commonly the point load index or unconfined compression test (UCS). There are other tests, such as the cone indenter or Schmidt hammer. Only the UCS test directly measures the compressive strength of the rock. All the other tests measure the strength indirectly, usually involving tensile failure to enable the test equipment to be lighter and therefore portable and suitable for field use.

The rock strength table given in Table 5.7 is after EN ISO 14689-1 and follows the practice and terminology outlined in ISRM suggested methods. These strength classes differ significantly from the previous BS terminology as shown in Table 5.8. The changes are such that at the weaker end of the range, the definitions of terms are a whole category different. Thus a weak rock which was between 1.25 and 5 MPa is now between 5 and 25 MPa. This is a very significant difference. Loggers and their

Table 5.8 Comparison of rock strength terms used in UK practice

Term	2007	1999
Extremely weak	0.6–1.0 in UK	Not a defined term
Very weak	1–5	<1.25
Weak	5–25	1.25–5
Medium strong	25–50	5–12.5
(Moderately strong) term discarded		12.5–50
Strong	50–100	50–100
Very strong	100–250	100–200
Extremely strong	>250	>200

TIP ON CHECKING FIELD STRENGTH DESCRIPTOR AGAINST TEST RESULTS

- The strength assessed by the logger in the field is the material strength between the joints and therefore the highest strength available.
- Tests to measure strength involve larger samples in which the presence of defects will reduce the measured strength.
- The different test methods (such as vane, triaxial, unconfined compression, point load, cone indenter, Schmidt hammer) each test the soil or rock in different ways and at different strain rates.
- It is therefore not surprising that these various tests tend to give different answers.
- As long as it is confirmed that the logger is correctly calibrated, the field descriptions should remain unchanged by the results of other strength tests. Any apparent inconsistencies in assessed and measured strengths should be an issue for discussion in the ground investigation report (GIR).

supervisors should take great care to know which standard and set of definitions they are following on any project. The field record sheets and the final field logs should clearly state which approach has been in use in the field logging.

It can be seen that the term 'moderately strong' has been lost in the above change. Whilst there is no particular attraction to this term, the loss of the boundary at 12.5 MPa is significant. Rocks of 5 MPa are only just about testable in any laboratory, can be broken by hand and are still weak rocks to most practitioners. Rock with a compressive strength of 25 MPa is equivalent to a weak concrete and thus is quite a competent material. This range of behaviour is too wide for a single class and it is suggested that this boundary should be reinstated. This is a particular problem as many other standards and text books use this boundary in terms of presumed bearing capacities in rocks and this relevant information will now not be available on field logs.

Combination of the terms used in the description of the consistency and strength of soils and rocks is given in Table 5.9. This has been compiled from BS 5930 and EN ISO 14689-1 and also incorporates the guidance incorporated within the geological strength index (GSI) presented by Marinos and Hoek (2000).

Table 5.9 Combined descriptive terms for soil consistency and strength and rock strength

Term for field use	Field estimate of soil consistency or rock strength	Examples	Strength
Extremely soft	Finger pushed in easily to full extent; unable to maintain shape.		Extremely low c_u <10 kPa[1]
Very soft	Finger pushed in up to 25 mm; exudes between fingers.	Recent normally consolidated estuarine, lacustrine or alluvial clay	Very low c_u = 10–20 kPa
Soft	Finger pushed in up to 10 mm; moulds by light finger pressure.	Organic soil	Low c_u = 20–40 kPa
Firm	Thumb makes impression easily; cannot be moulded by fingers; rolls to thread.		Medium strength c_u = 40–75 kPa
Stiff	Indented slightly by thumb; crumbles, breaks; remoulds to lump.	Desiccated clay, weathered overconsolidated clay, glacial till	High c_u = 75–150 kPa
Very stiff[3]	Indented by thumbnail; crumbles, does not remould.	Overconsolidated clay, argillaceous mudrock, mudstone weathered to clay	Very high c_u = 150–300 kPa
Hard[3,4]	Scratched by thumbnail; brittle behaviour.	Glacial till	Extremely high c_u >300 kPa
Extremely weak[3,4]	Scratched by thumbnail; gravel size lumps crush between finger and thumb.	Stiff fault gouge, destructured mudstone	q_u = 0.6–1 MPa[2]
Very weak	Gravel size lumps can be broken in half by heavy hand pressure; crumbles under firm blows with point of a geological hammer; can be peeled by a pocket knife.	Distinctly weathered or altered rock, shale	q_u = 1–5 MPa

Weak	Can be peeled with a pocket knife with difficulty; shallow indentation made by firm blow with point of a geological hammer; rock broken by hammer blows when sample held in hand; thin slabs, corners or edges can be broken off with heavy hand pressure.	Chalk, claystone, potash, marl, siltstone, shale, rocksalt	$q_u = 5{-}25$ MPa
Medium strong	Cannot be scraped or peeled with a pocket knife; specimen resting on a solid surface can be fractured with a single blow from a geological hammer.	Concrete, phyllite, schist, siltstone	$q_u = 25{-}50$ MPa
Strong	Specimen resting on a solid surface requires more than one blow of a geological hammer to fracture it.	Limestone, marble, sandstone, schist	$q_u = 50{-}100$ MPa
Very strong	Specimen requires many blows of a geological hammer to fracture it; specimen chipped by hammer blows.	Sandstone, basalt, gabbro, gneiss, granodiorite, tuff	$q_u = 100{-}250$ MPa
Extremely strong	Specimen can only be chipped with a geological hammer; can only be broken by sledge hammer; rings on hammer blows.	Fresh basalt, flint, gneiss, granite, quartzite	$q_u > 250$ MPa

[1] c_u is the symbol for undrained shear strength.
[2] q_u the symbol for undrained compressive strength.
[3] For description of sands and sandstones at the soil–rock boundary, see Table 4.11.
[4] The term hard is reserved for use on transported soils; in situ deposits are described using rock terms.

6 Structure, Fabric and Texture

The internal structure of the soil or rock is very important in controlling the overall behaviour of the mass and a variety of interesting features come under this heading. For instance, the structural features change the orientation and continuity of the mass, fabric features change the mass permeability and thus rates of consolidation and textural features affect the stability of the mass for particular end uses. The presence and character of structure and fabric in rock and soil need to be described to provide a full representation of the material but the level of attention paid to the logging of these features is variable and tends to be low.

Structure, fabric and texture features are observable at three different scales, as shown in Figure 6.1. Structure is at the large scale and will generally only be observable directly in large field exposures. Fabric is at the medium scale and includes features that could be discernible and reportable in the scale of exposures normally available in field investigation such as a trial pit or borehole. Texture is at the small scale and includes those features which may require the use of a hand lens or microscope for

STRUCTURE – some visible in cores

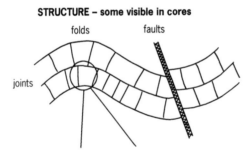

FABRIC – visible to the naked eye and in cores

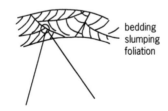

TEXTURE – possibly visible to the naked eye, often use lens or microscope

Figure 6.1 Scale of structure, fabric and texture features.

examination. The description of these features at whatever scale is generally carried out using standard geological terms as given in relevant texts and dictionaries. As these terms already fill a number of text books elsewhere, it is not practical or desirable to replicate this suite of terms and their definitions in this book.

6.1 Structure

Structure is the large scale interrelationship of individual bedding or structural units and discontinuities found in folds and faults. The scale of these features is such that their description is not usually possible in small exposures or cores and so would not generally appear on a field log. It may be possible to infer the presence of structure from features visible in a small exposure but full identification will require the assimilation of data from a number of exposures or sources of information.

It is important in all scales of field logging to record any information which may provide evidence of features not actually visible in the exposure being recorded. The presence of striated or polished surfaces, brecciated materials or small changes in dip could each indicate a structural feature in the vicinity of the borehole. All such features should be included in the description even if their potential association cannot be determined from the individual investigation hole being logged.

6.2 Fabric

Fabric comprises the medium scale features and includes changes in lithology, including aspects such as bedding, interbedding and jointing. Fabric can usually be seen in small exposures or cores and so can normally be included in most field descriptions. Fabric features in soils can include bedding, laminae, partings and lenses. Fabric terms in sedimentary rocks could include 'bedded' or 'laminated'; metamorphic rocks may be 'foliated' or 'banded'; igneous rocks may be 'flow-banded'. Fabric features also include the whole range of tensile or shear discontinuities such as joints, fissures or shears. The description of discontinuities is given in detail in Chapter 11.

There is a wide range of fabric features in soils and rocks which should be described such as those listed in Table 6.1. These features can each have a significant effect on the behaviour of the soil or rock, and information on the type, spacing and orientation is vital to allow realistic interpretation of test results and thus the assessment of the overall behaviour of the ground.

The description of features classed here as inhomogeneities should include information on the composition, size and spacing. The range of fabric features is shown diagrammatically in Figure 6.2. The important feature to note from this figure is that bedding features have thickness and spacing and these both need to be described, so giving 'medium spaced thin beds' for example. Discontinuities have spacing which should be described but no thickness, hence 'extremely closely fissured' for example. Partings are a special case of bedding features which have spacing but no thickness, being only a grain or two thick and so would be described for example as 'very closely spaced partings'. The presence of fabric and discontinuities should be noted in the first sentence of the description with the consistency descriptor, e.g. 'stiff fissured', 'extremely weak sheared', 'laminated' or 'closely jointed'. Additional information on the character of the fabric features and discontinuities would follow in subsequent sentences. Details of any discontinuities

Table 6.1 Types of fabric to be described

Fabric	Example	Property affected
Bedding	Partings Laminae Beds Lenses	Homogeneity
Tensile discontinuities	Fissures Desiccation cracks Joints	Strength
Sheared discontinuities	Shear surfaces	Strength
Other discontinuities	Bedding structures Root tracks Lithorelicts Bioturbation Cryoturbation Anisotropy	Homogeneity

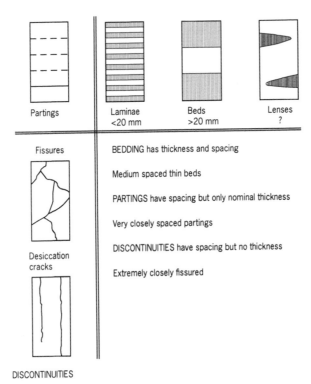

Figure 6.2 Bedding and discontinuity fabrics and the relationship between thickness and spacing.

should include information on the orientation, spacing, roughness, aperture and any other items of information from Chapter 11.

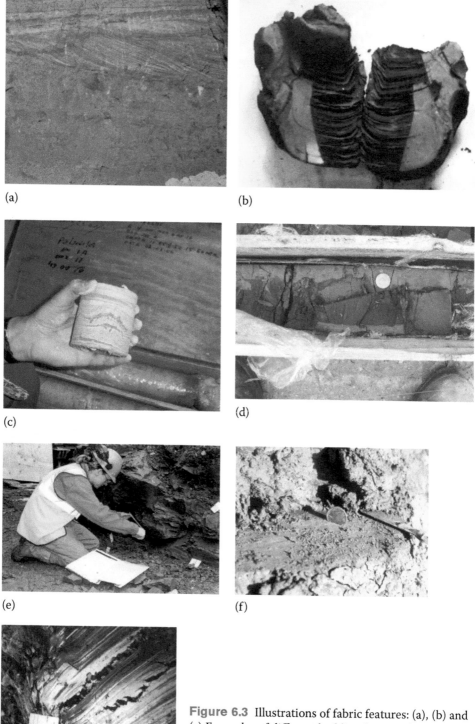

(a)

(b)

(c)

(d)

(e)

(f)

(g)

Figure 6.3 Illustrations of fabric features: (a), (b) and (c) Examples of different bedding features (photograph (c) courtesy of Steve Macklin); (d) and (e) examples of fissuring in core and field exposure; and (f) and (g) examples of shear surfaces: (f) a periglacial shear in weathered mudstone and (g) a shear in mudstone fill.

TIP ON FABRIC LOGGING

Observation of fabric features is often very limited when the exposure is freshly opened with interbeds or discontinuities being very difficult to see.

- Fabric logging should initially be carried out on the fresh exposure.
- The exposed material should then be left to air dry overnight and the logged record revisited. This may need to be repeated more than once.
- Air drying allows the coarser materials such as fine sand or silt to dry to a lighter colour and for tensile discontinuities to pull apart slightly as the fissure bound blocks shrink slightly.
- A longer period of air drying may be required to expose all features. This could apply particularly to shear surfaces. Experience has shown that it can take several weeks for the drying to advance sufficiently for the two sides of the shear to pull apart.
- Photographic records should be taken of all fabric visible to support detailed fabric logs.
- Examples of some fabric features are illustrated in Figure 6.3 and some examples of fabric descriptions in accordance with the rules outlined above follow.

EXAMPLE FABRIC DESCRIPTIONS

- CLAY with closely spaced lenses (5 by 10 to 15 by 25 mm) of peat.
- (NB The terms 'lens' and 'pocket' have no size connotation so dimensions should always be given.)
- thickly interlaminated SAND and CLAY;
- CLAY with closely spaced thin laminae of sand;
- CLAY with extremely closely spaced fine sand partings;
- firm fissured and sheared CLAY. Fissures are generally subvertical very closely spaced smooth planar grey gleyed. Shears are subhorizontal (175/05 to 200/10) up to 500 mm persistence smooth undulating straight striated highly polished.
- very closely jointed very thinly bedded SANDSTONE. Joints are 45–60° undulating rough tight and clean.
- extremely closely sheared laminated MUDSTONE. Shears dip up to 5–10° planar smooth polished and lightly striated.

BACKGROUND ON BEDDING NOMENCLATURE

- A bedding plane is not necessarily a discontinuity in the sense of a mechanical break in the core.
- It is perfectly feasible to recover laminated rocks in core sticks of substantial length, such as several metres.
- The record of whether the bedding plane is a mechanical break in the core will be picked up in the description of the discontinuities as bedding fractures.
- The thickness of the bedding units and the spacing of the fractures need not be the same measurement.

The description of the bedding in the above examples is based on the basic geological definition which is that of a depositional surface. The thickness or spacing terms given in Table 6.2 are used as appropriate. Bedding planes can arise from a variety of field situations including pauses in sedimentation, changes in material being deposited or changes in chemistry and thus of colour. The description of the

Table 6.2 Terms for description of spacing and thickness of fabric features

Typical thickness or spacing (mm)	Bedding term	Discontinuity term
over 2000	Very thickly bedded	Very widely spaced
2000–600	Thickly bedded	Widely spaced
600–200	Medium bedded	Medium spaced
200–60	Thinly bedded	Closely spaced
60–20	Very thinly bedded	Very closely spaced
20–6	Thickly laminated	Extremely closely spaced
under 6[1]	Thinly laminated	Extremely closely spaced

[1]The thinnest bedding or closest spacing terms represent dimensions that are quite coarse. Additional refinement can be provided by providing measurements where these are less than 20 or 6 mm.

TIP ON DESCRIPTION OF DISCONTINUITY SPACING
- The presentation of descriptions with too wide a range of spacing or thickness provided by the terms used is not helpful.
- It is better if, in addition to the use of the descriptors which encompass the range of most common spacing, the actual range of spacing or thickness is given as a measurement. Measurements given can clearly include the full range of spacing or thickness that has been observed. This approach is shown below.
- It is good and normal practice to report the spacing as a minimum/mode/maximum in mm, for example (25/65/130).
- This dual approach allows the descriptor to be really helpful, without any concerns arising from not including the extremes.

EXAMPLE DESCRIPTIONS
Noting the different ranges indicated by the term and the measurements:
- very closely fissured (15/40/65)
- generally closely spaced desiccation cracks (25/100/220)
- very thinly bedded (15/50/75).

---

bedding should record the bedding features seen, including their nature and the thickness of the bedding units they define.

The spacing is measured normal to the discontinuity surface and not along the scan line such as a core axis. The thickness and spacing descriptors should be given for the most common range present using terms from Table 6.2 and do not necessarily need to cover the whole range just in order to include rare extremes.

6.3 Texture

The texture of a rock refers to the small scale features of and between individual grains or agglomerates of grains. These grains may show a preferred orientation or varying bonding, cemented or interlocked relationships. Terms used to describe texture might include normal geological terms such as 'porphyritic', 'crystalline', 'cryptocrystalline', 'amorphous' and 'glassy'. Examination of the texture may require the use of a hand lens or the microscopic examination of a thin section of soil grains or the rock. These sorts of textural descriptions are usually prepared by a specialist subsequent to the field logging and use small subsamples or hand specimens. In fact they are often carried out as part of a laboratory testing programme, such as petrographic descriptions. For this reason, detailed textural descriptions may appear in a field log but are more commonly presented as some form of attachment to the investigation report.

7 Colour

Description of colour in soils and rocks is always required. Although the colour is not an intrinsic engineering property of the ground, colour can be critical to the assessment of the ground conditions by helping to identify geological processes that have affected the materials or through information on the position in the stratigraphical sequence at the site. The presence of particular colours can suggest certain significant chemical or mineralogical conditions (e.g. dark colours indicating organic content, mottled colours indicating chemical weathering) or potential processes (e.g. red laminated shale indicating burning, light grey gleying indicating reductive processes which can suggest possible shearing).

A simple terminology is generally sufficient for the description of colours and this is outlined in Table 7.1.

This is not the complete list of colours that can be used, but these terms should cover a very large majority of logging requirements. Colour terms that are ambiguous or cover a range of colours should not be used. Examples of such colours include olives which vary from green to black, charcoal can range between grey and black and lilac varies from white through yellow to purple.

Many soils present with more than one colour and these should be described. These colours may appear as mottling, banding or associated with different constituent materials; examples are shown in Figure 7.1. A variety of influences during soil formation result in the colours becoming blotchy or mottled and these can include depositional, pedogenic and weathering processes. Chapter 10 provides further information on what colours might be seen arising from weathering

Table 7.1 Terms for description of colour

Lightness	Chroma	Hue
Light	Pinkish	Pink
Dark	Reddish	Red
	Yellowish	Yellow
	Orangish	Orange
	Brownish	Brown
	Greenish	Green
	Bluish	Blue
		White
	Greyish	Grey
		Black
Additional hues which can be found in some rocks		Purple
		Cream

TIP ON COLOUR DESCRIPTION

There are some simple rules to be followed in the description of colour:
- Keep it simple and give the overall effect. This requires the logger to stand back from the exposure rather than examining all the colours too closely.
- The colour is to be observed when the soil or rock is damp. Accurate colour description will often require the sample or core or exposure to be wetted up. A hand held garden sprayer is ideal for this purpose.
- Colours can come in certain patterns as discussed below and these possible patterns include:
  - mottled when colours are arranged in irregular spots or blotches
  - multicoloured when there are more than three distinct colours
  - banded where changes may be related to bedding or differences in materials present.
- Note changes that occur after the sample is opened or the exposure created. Such changes may be due to, say, moisture content or oxidation. Changes caused by the exposure drying out are not usually worth recording.

(a)

(b)

(c)

Figure 7.1 Examples of coloured soils and rocks: (a) mottling in very stiff clay; (b) colour banding in extremely weak mudstone; and (c) black gravel with green sandy clay matrix.

and how they arise. The description of such irregular colouring, termed 'mottling', where present in up to three colours can helpfully be made as this provides useful evidence to indicate the origin and history of the soil. For instance, a weathered mudrock might be 'yellowish brown, brown and light grey mottled' reflecting the combined effects of aerobic and anoxic conditions that obtained during the weathering under the various stages of periglacial conditions.

The description of colour mottling should record the range of colours present and a form of words, which is largely common sense, is indicated in Figure 7.2. The mottling description can indicate the dominant colours by adjusting the word order.

It is suggested that the description of more than three colours in any soil or rock is unlikely to be helpful. Colour descriptions should give an overall impression of the soil or rock, not a detailed, close up list of the many colours that might be present, say in a terrace gravel (the highest number seen is 13 colours in a gravel from the River Lee in Cork, Ireland). This is unlikely to be meaningful and just hides the more important details from the reader. In cases where more than three colours are present, the use of the term 'multicoloured' is appropriate.

It is also possible for the coarse (clasts) and fine (matrix) fractions of a soil to present different colours. In these instances the colour of both fractions should be provided. An example of such a soil might thus be 'green sandy clay and black flint gravel' as shown in Figure 7.1.

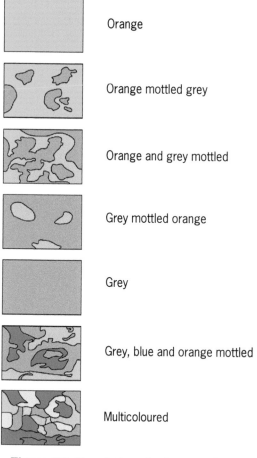

Figure 7.2 Description of colour mottling.

TIP ON MULTICOLOURED SOILS AND ROCKS
- There are certain horizons where the use of the term 'multicoloured' can obscure important information.
- For example, the mottled beds within the Lambeth Group (Lower Tertiary deposits in the London Basin) include several layers with more than three colours, but distinction needs to be drawn between those where the white and light grey colour assemblage indicates the presence of calcrete in contrast to those where a variety of browns indicates different concentrations of ferric and ferrous compounds in the soil. The first is likely to be significant for excavation and water flow in most engineering works, whereas the latter is unlikely to be so important.
- This important distinction would be lost if both horizons were simply described as multicoloured.

Some examples of colour descriptions using this terminology and following the above rules are given below.

EXAMPLE DESCRIPTIONS
- light brown CLAY
- reddish brown SANDSTONE
- dark bluish grey laminated MUDSTONE
- white quartzitic SANDSTONE
- grey mottled orangish brown slightly fine sandy CLAY
- grey, light grey and reddish brown mottled CLAY
- multicoloured GRAVEL and COBBLES
- reddish brown and grey thinly colour banded slightly sandy CLAY
- black oxidising to dark brown fine sandy organic SILT.

Consistency of colour description is more important than absolute accuracy. Colour charts can be used to help and these come in a number of forms.

An industry standard for colour description is the Munsell colour chart which was developed by the US Soil Conservation Service for classifying the colour of soils; the charts can also be used for rocks. The use of Munsell colour charts as called for in ISO 25177: 2008 on the description of agricultural soils is advocated by many such as Smart (2008). The origin and use of Munsell charts is based in the description of soils for soil science and taxonomic purposes such as are used in national soil surveys and soil quality assessments.

These charts are available from a number of suppliers which can be found on the internet. They provide a standard set of colours and are therefore favoured by some in industry. Indeed, whether such charts should be mandatory or not is an issue still under debate. Those who favour their use cite their origin based in soil quality description and the removal of all possibility of subjective description. On the other hand, their use is impractical because the time taken to find the correct colour amongst the over 300 presented accords too much time and creates an over-emphasis on this single element of preparing a soil or rock description. In addition, the descriptors provided from these charts are meaningless to any reader using the English language as they are presented in an alphanumerical code with words.

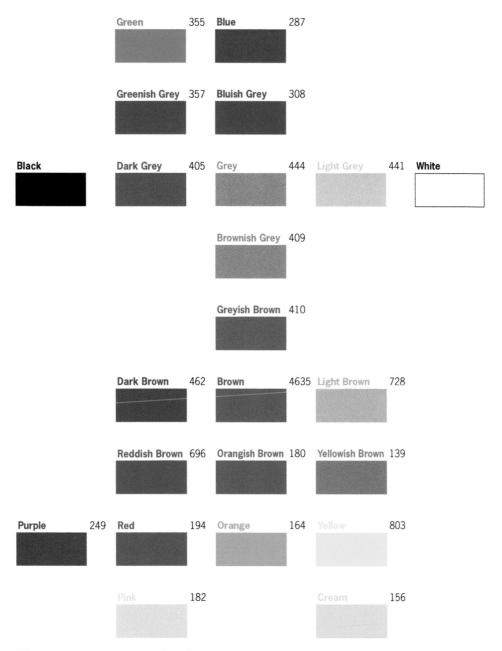

Figure 7.3 Simple colour chart for general use, to be read in conjunction with Table 7.2. The above are examples of Pantone colours from Pantone® colour charts.

The preference in this book is for a simple colour chart such as that presented in Figure 7.3. Experience has shown that this simple chart, which uses normal colour terms which are comprehensible to the ordinary practitioner, should suffice for the description of colour for a large majority of soils and rocks in practice. Figure 7.3 includes the relevant Pantone codes in order to offer some guidance and attempt at some standardisation in colour terms as applied in practice. Translation from the Pantone colours to the equivalent R/G/B and hexadecimal colour codes is provided in Table 7.2.

It should be noted that the colours that are output by various office printers today do vary and so this sheet needs to be used with care and an element of subjectivity necessarily comes into the matter. Although the variance is not great, it can be significant, but in general a laser printer may be better than an inkjet printer. In any case printing at a higher resolution produces a flatter tone and therefore better chart and printing on better quality paper is helpful. The best seems to be a matt photo paper, but this may not be practical for all users.

There is no reason why individuals cannot adjust the colours chosen for Figure 7.3 to provide colours that they prefer, but the recording of colour should be consistent during the life of a project and particularly between different practitioners on a project. This is usually more important than there being a need for great accuracy in absolute terms. There are a small number of geological formations where the accurate description of colour is important in order to help in correctly determining the geological succession within the unit. A very good example of where this is considered to matter is the Lambeth Group in the London Basin.

In addition, and as a factor that is under our control, the description of colour is affected by the lighting at the point of logging, be this out in the field or in the

Table 7.2 Colour conversion codes

Colour	Pantone as given in Figure 7.2	Red/green/blue	Hexadecimal
Black			
Dark grey	405	107/94/79	#6B5E4F
Grey	444	137/142/140	#898E8C
Light grey	441	209/209/198	#D1D1C6
White			
Brownish grey	409	153/137/124	#99897C
Greyish brown	410	124/109/99	#7C6D63
Dark brown	462	91/71/35	#5B4723
Brown	4635	140/89/51	#8C5933
Light brown	728	211/168/124	#D3A87C
Blue	287	0/56/147	#003893
Bluish grey	308	0/96/124	#00607C
Green	355	0/158/73	#009E49
Greenish grey	357	33/91/51	#215B33
Reddish brown	696	142/71/73	#8E4749
Orangish brown	180	163/68/2	#A34402
Yellowish brown	139	175/117/5	#AF7505
Purple	249	127/40/96	#7F2860
Red	194	153/33/53	#992135
Orange	164	255/127/30	#FF7F1E
Yellow	803	255/237/56	#FFED38
Pink	182	249/191/193	#F9BFC1
Cream	156	242/198/140	#F2C68C

TIP ON COLOUR CHARTS

- Although commercial colour charts are available, the simplest and cheapest are paint or other commercial colour swatches available from the local DIY store. Charts of the colours can be made up and issued to each logger on the site in order to ensure that the descriptions are consistent.
- The colour terms given on the paint charts should not be used under any circumstances.

BACKGROUND ON COLOUR BLINDNESS

- It is not generally recognised how much the description of colour is complicated by the high proportion of colour blindness in the population. The most common form is red-green colour blindness. Because these defects are inherited as recessive traits, the incidences are much higher in males than in females. Incidences of red-green colour deficiencies also vary between human populations of different racial origin, with the highest rates being found in Europeans and the Brahmins of India (about 8% of males) and Asians (about 4%).
- If loggers are aware of their colour blindness, the use of colour charts, at least in their area of difficulty, should be mandatory.

logging shed. Logging carried out in the open air should encourage the description to be made under bright cloudy conditions. Loggers should try to avoid direct sunlight which blanches colour, or unduly gloomy days which subdue colours and enhance greyness. Most indoor logging areas are provided with various forms of fluorescent lighting and this usually gives an orange or yellow hue to the light. Logging areas should be lit by blue or daylight bulbs (CIE D65 which represents noon daylight (6504K), or CIE C which represents an average northern daylight (6774K)) and dedicated description areas should be fitted with such bulbs as a matter of course.

The ideal indoor logging area is benches under large windows facing north (in the northern hemisphere) with plenty of light fittings over the work area fitted with daylight bulbs to extend the working hours on gloomy days or in winter.

In training laboratory technicians in colour description, it has been found that provision of colour charts such as that in Figure 7.3 is of great assistance in making descriptions more uniform and much less dependent on the area of the laboratory where the work is carried out. This helps minimise variances in colour description between operators and between different areas of the laboratory where lighting conditions can normally be expected to vary.

8 Secondary and Tertiary Fractions

8.1 Secondary fractions

Having identified and named the principal soil type, this NAME should normally appear in capitals in the completed description so as to be clearly visible. The next step is to identify and describe the secondary fractions. To remind readers, the secondary constituents are those that modify the behaviour of the dominant principal fraction.

There is a range of terms which are used to describe the secondary fractions and the terms that are to be used depend on the principal soil type. This range of terms is set out below. It is important to note that in the assessment of the proportion of secondary fractions, each fraction is assessed individually. Thus in a 'very gravelly slightly clayey SAND' the proportions of the gravel and clay are assessed and reported separately. This is logically necessary as combining all the secondary constituents into a single descriptive category would lose important detail. This logical position gives rise to the possibility that a reported soil description could have more than 100% constituents if due care in assembling the description is not given. A soil can only be made up of 100% constituents.

However, this rule needs to be adjusted when a soil includes very coarse particles. In this instance, the very coarse fraction and the combined coarse and fine fraction are each described as 100% soils before being combined. The combined soil can therefore appear to be more than 100%. The approach to dealing with this issue is given below.

8.1.1 Secondary fractions in very coarse soils

When the soil is made up entirely of very coarse particles, the terms in Table 8.1 are used to describe the relative proportions of the boulder and cobble particles present. Before carrying out this description, the cautionary comments in Chapter 4.4 about sample size should be borne in mind. The size of sample or exposure that should be available to the logger in order to describe these soils should be of the order of a tonne because of the presence of boulders. Even when the logger is presented with sufficient material for description, the application of the terms given below is far from easy. The availability of sufficient material may be possible in an open excavation or trial pit, but is certainly not possible in a borehole. Fortunately true mixtures of very coarse particles are not particularly common in Nature so this is not an everyday problem. It would therefore not be very common to see the terms in Table 8.1 appearing on field logs.

A more common circumstance is where the very coarse particles are mixed with some finer particles. In this case a different set of terms is used, as given in Table 8.2.

Table 8.1 Terms for description of very coarse soil mixtures

Term	Secondary constituent by weight
BOULDERS with occasional cobbles	up to 5% cobbles
BOULDERS with some cobbles	5% to 20% cobbles
BOULDERS with many cobbles	20% to 50% cobbles
COBBLES with many boulders	20% to 50% boulders
COBBLES with some boulders	5% to 20% boulders
COBBLES with occasional boulders	up to 5% boulders

Table 8.2 Terms for description of finer soil fractions in very coarse soil

Term	Secondary constituent by weight
Very coarse soil (BOULDERS and/or COBBLES) with a little finer material	up to 5% finer material
Very coarse soil (BOULDERS and/or COBBLES) with some finer material	5% to 20% finer material
Very coarse soil (BOULDERS and/or COBBLES) with much finer material	20% to 50% finer material

The 'finer material' indicated here relates to the coarse and fine fractions of the soil as distinct from the very coarse fraction. The coarse and fine fractions are described together as either a coarse or a fine soil in accordance with the procedures set out in Chapter 4 while ignoring the very coarse fraction. This separation into description of two fractions is in accordance with the process given in the flow chart in Figure 4.3.

The separate description of the very coarse particles as a 100% soil and the finer material as a 100% soil gives two complete soils. This is rendered back to a single soil by combining the two descriptions in the terms in Table 8.2 and illustrated by the examples in Table 8.3.

In practice the description is unlikely to be sufficiently precise for the above examples actually to become an issue for such soils, given all the problems in logging such materials. In particular the sample being observed or tested in the laboratory is unlikely to be large enough to be truly representative. In the unlikely event that this is actually critical in a particular investigation, the logger can derive a site specific procedure and classification. This should be presented to the reader along with the field log. It is always possible, of course, to carry out particle size analysis of very coarse soils, but this would usually be a field rather than a laboratory test.

The principal soil type name of the finer material may also be given in capital letters, e.g. 'BOULDERS with a little sandy GRAVEL', 'COBBLES with some sandy CLAY'. It is apparent that descriptions of such materials may often benefit from the multiple sentence approach, but not in every case. The first sentence should always include the quantification of all the fractions, as in these examples. The subsequent sentences can include detailed information on each of the fractions in turn as necessary.

Table 8.3 Examples of very coarse soil descriptions including finer material

Case	Description	Example grading
Very coarse materials only	COBBLES with many boulders	• 70% cobbles, 30% boulders
Very coarse material with finer material	COBBLES with some fine sandy CLAY	• 85% cobbles, 15% finer material • Finer material is 40% sand, 35% silt, 25% clay • Overall grading if comparing with a particle size analysis of 85% cobbles, 6% sand, 5% silt and 4% clay
Mixed very coarse materials with finer material	COBBLES with some boulders and much very sandy GRAVEL	• Very coarse fraction makes up 65% plus 35% finer material • Very coarse fraction is 85% cobbles, 15% boulders • Finer material is 70% gravel, 30% sand • Overall grading if comparing with a particle size analysis of 55% cobbles, 10% boulders and 25% gravel, 10% sand

EXAMPLE DESCRIPTIONS
- Generally rounded COBBLES of medium strong limestone with some firm brown sandy CLAY.
- BOULDERS with some cobbles and much sandy GRAVEL. Boulders and cobbles are up to 300 mm of well rounded to sub rounded strong sandstone. Gravel is angular quartzite.

8.1.2 Very coarse particles as a secondary fraction

The more common occurrence of very coarse particles will be as a secondary fraction within a finer material, most commonly a coarse soil. The terms in Table 8.4 are used for the description of this mixture with the boulder and cobble fractions being assessed and reported separately. As before, the finer material comprises the coarse and fine soil fractions.

Following the procedure outlined above and in the standards (see Figure 4.2) the finer fraction is described as a 100% soil in its own right. As in very coarse soils,

Table 8.4 Terms for description of very coarse fractions in finer soil

Term	Secondary constituent by weight (%)	Term	Secondary constituent by weight (%)
FINER MATERIAL with high boulder content	>20	FINER MATERIAL with high cobble content	>20
FINER MATERIAL with medium boulder content	5–20	FINER MATERIAL with medium cobble content	10–20
FINER MATERIAL with low boulder content	<5	FINER MATERIAL with low cobble content	<10

Table 8.5 Examples of finer material descriptions including very coarse soil

Case	Description	Actual grading
Finer material with very coarse	Very sandy very clayey GRAVEL with low boulder and medium cobble content.	• Finer material is 45% gravel, 25% sand, 15% silt and 15% clay. • To this 100% soil should be added 3% boulders as well as 15% cobbles. • Overall grading if comparing with a particle size analysis of 37% gravel, 21% sand, 12% silt, 12% clay plus the 3% boulders and 15% cobbles.
Finer material with unrepresentative sample of very coarse	Clayey sandy GRAVEL with one boulder. At 6.5–6.9 m driller reports pushing boulder and chiselling. Angular coarse gravel size sandstone fragments recovered.	Actual grading not reliably assessed because not enough material seen or recovered.

the proportions of the cobbles and boulders are individually assessed and these will be in addition to the finer material. This procedure is illustrated in Table 8.5 for the two common cases of a coarse soil containing very coarse particles. It is important to be very clear about this principle in order to ensure that the description of the principal soil type, the coarse soil in this case, and the secondary constituents is correct.

Although obtaining a representative amount of sample for the description is likely to be more practical in a coarse soil, the size of sample remains a problem. Trial pit or other excavations will probably provide sufficient material for the logger to make a reasonable assessment of the soil's grading and thus to prepare an accurate description. However, this remains impossible in boreholes, as both insufficient material will be available and very coarse particles cannot be recovered in any case. In these cases it is much better and strongly recommended that all information relating to very coarse particles should be presented on the field log as shown in Table 8.5. This information should include any observations by the driller, the chiselling records and notes on the careful observation of the shape of the particles recovered, in order that the reader can make their own assessment of this size fraction and its significance to the works.

The details of the coarse and very coarse particles that are included in a description vary and also an inconsistent word order is commonly used in the description. This is not the fault of the logger, as guidance is not provided in the Standards about what should be included. However, even this apparently plain remark is ambiguous. A suggestion is put forward in Table 8.6 to tidy up this area of descriptive practice.

TIP ON TERMINOLOGY FOR MULTIPLE LITHOLOGIES
- When describing the lithologies of coarse soils, the terms 'various lithologies' or 'mixed lithologies' mean more than two are present and are satisfactory for use.
- 'Assorted lithologies' should not be used, as the term could be confused with a comment on the shape of the grading curve and so is potentially ambiguous.

Table 8.6 Details to include in describing very coarse and coarse soils

Size fraction		Preferred word order and example description
BOULDERS	Details to include	**Dimensions, shape, strength, lithology**
	• example as principal	200–400 mm rounded BOULDERS of strong basalt
	• example as secondary	Boulders are 200–400 mm rounded of strong basalt.
COBBLES	Details to include:	**Shape, strength, lithology**
	• example as principal	Subangular COBBLES of medium strong limestone
	• example as secondary	Cobbles are subangular of medium strong limestone.
GRAVEL	Details to include:	**Shape, grading, lithology**
	• example as principal	Subrounded fine to coarse sandstone GRAVEL
	• example as secondary	Gravel is subrounded fine to coarse of sandstone.
SAND	Details to include:	**Grading, lithology** when not quartz
	• example as principal	Fine occasionally coarse shelly SAND
	• example as secondary	No detail required.

Where GRAVEL is the principal soil type description of the shape, grading and lithology should normally be made in accordance with Table 8.6. However, where gravel is a secondary constituent, description of the shape, grading and/or lithology is normally warranted only where it contributes to the understanding of the deposit.

Similarly, where SAND is the principal soil type, description of the grading should be made. Description of the lithology is not given unless it is of particular interest and that is usually when it is of a material other than quartz (or any related material such as flint or chert). Where sand is a secondary constituent, description of the grading is not normally warranted.

Multiple sentences can be used in the description of coarse soils where required to avoid confusion and ambiguity as outlined in Chapter 3.2. However, this should not often be required in coarse soils and current practice is to overuse this approach. A second sentence is generally only appropriate when detail is required on a secondary constituent.

EXAMPLE DESCRIPTIONS
- Slightly medium to coarse sandy subrounded fine GRAVEL of medium grained sandstone;
- Very gravelly fine SAND. Gravel is subangular fine and medium of limestone;
- Slightly sandy fine to coarse GRAVEL with occasional cobbles. Gravel and cobbles are subrounded of various lithologies;
- Grey and light grey very medium to coarse sandy subangular GRAVEL of tabular fine and medium weak mudstone.

8.1.3 Secondary fractions in coarse soils

The terms for the description of coarse soils alone, that is mixtures of gravel and sand, are given in Table 8.7. This should be a reasonably straightforward exercise. The particles are recoverable from boreholes and excavations, representative

Table 8.7 Terms for description of mixed coarse soils

Secondary constituent by weight	Term
up to 5%	slightly sandy GRAVEL OR slightly gravelly SAND
5% to 20%	sandy GRAVEL OR gravelly SAND
20% to 40%	very sandy GRAVEL OR very gravelly SAND
about 50%	SAND AND GRAVEL

samples are of modest size (less than 20 kg) and the particles can easily be handled, measured and inspected.

The descriptive classes given in Table 8.7 are illustrated in Figure 8.1 which shows a suite of sieved coarse soil mixtures complete and then with the sand and gravel fractions separated.

The field description of the particle size distribution of a soil faces the difficulty that the description has to be made in accordance with the percentages by weight as this is the measurement made in the laboratory particle size analysis. It is thus the case that the logger has to override the eye and brain which are seeing and assessing the percentages by volume. This requires a mental translation to be made to allow for the different mass of a volume of loose particles of various sizes and the particle density of the coarsest particles. This can be done using the typical densities suggested in Table 8.8. The lowest density material is the pile of loose sand in the hand or on the bench and the descriptive adjustment is made from this datum. The description of particles in coarse and fine soils needs to be considered separately:

- In coarse soils the logger is assessing the proportion of gravel and cobble size pieces within the sand 'matrix'. Using the typical densities given, the ratio of the density of the loose sand particles to the density of loose gravel particles is about 0.75. The ratio of the density of loose sand particles to that of individual cobble or gravel particles is 0.6. The logger cannot see the fine soil particles and so the proportion of clay and silt is assessed by hand, using the hand tests given in Chapter 4.8.
- In fine soils the ratio of the density of the clay matrix to the individual cobble or gravel particles is about 0.8. The logger cannot see the sand particles and so the proportion of sand is assessed by feel.

Using these densities it is possible to show the appearance of the various defined proportions by mass of the secondary constituents in volume terms. The sand/gravel ratios of the densities given above fall in the range 0.6 to 0.8; whilst this appears to be a large range, field description is not a precise science and so it is reasonable to assume an average ratio of 0.7. For the purposes of illustration it is also reasonable to assume that the proportion of particles that can be seen in a two dimensional view can be taken to be the same as the volume in three dimensions. The classes of secondary constituent descriptors (Table 8.7) are thus illustrated together with photographs of the sieved samples in Figure 8.1 and Figure 8.2 for coarse and fine soils, respectively.

COARSE SOIL

Composite soil = sand + gravel fractions

Slightly gravelly SAND   Gravel 4% Sand 96%

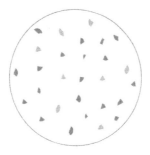

5% gravel particles by mass in sand   Proportion by volume = 3.5%

Composite soil = sand + gravel fractions

Gravelly SAND   Gravel 12%, Sand 88%

Figure 8.1 Chart for volume percentage estimation of coarse soils, assuming a density adjustment ratio of 0.7. Chart shows a suite of sieved coarse soil mixtures complete and then with the sand and gravel fractions separated. The diagrams represent the boundary between each class, as defined in Table 8.7 whilst the photographs are within each class and the actual proportions are given (photographs courtesy of Soil Mechanics).

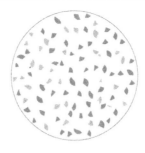

20% gravel particles by mass in sand   Proportion by volume = 14%

Composite soil = sand + gravel fractions

Very gravelly SAND   Gravel 30%, Sand 70%

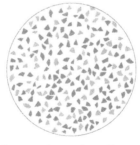

40% gravel particles by mass in sand   Proportion by volume = 28%

Composite soil = sand + gravel fractions

SAND AND GRAVEL   Sand 50%, Gravel 50%

Figure 8.1 *Continued*

60% gravel particles by mass in sand   Proportion by volume = 42%

Composite soil = gravel + sand fractions

Very sandy GRAVEL   Sand 30%, Gravel 70%

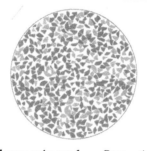

80% gravel particles by mass in sand   Proportion by volume = 56%

Composite soil = sand + gravel fractions

Sandy GRAVEL   Sand 12%, Gravel 88%

Figure 8.1 *Continued*

95% gravel particles by mass in sand   Proportion by volume = 66.5%

Composite soil = sand + gravel fractions

Slightly sandy GRAVEL   Sand 4%, Gravel 96%

Figure 8.1 *Continued*

8.1.4 Fine soil as a secondary constituent

When the coarse soil also includes fine soil constituents the terms used to describe this secondary fraction are as given in Table 8.9. The percentages are the same as for coarse secondary constituents, but the assessment of the grading is more difficult as this relies on assessing the proportion by feel in a material that is essentially gritty.

Using the same approach as above and an adjustment factor of 0.7, the appearance in volume terms of the mass proportions of fine and coarse soil mixtures is shown in Figure 8.2.

One of the notable contrasts between Figure 8.1 and Figure 8.2 is between the coarse soil with 30% gravel which is 'very gravelly SAND' and a fine soil with 30% gravel which is only 'slightly gravelly CLAY' reflecting the dominant effect of proportions of fine soil, particularly clay.

In practice in the field, each of the three classes of slightly clayey, clayey or very clayey are assessed in terms of how it behaves when being moulded in the hand and how dirty it makes your hands, as follows:

- Slightly clayey secondary component will not really make the hand dirty.
- Clayey secondary component will make the hand dirty.

Table 8.8 Density ratios in disturbed samples

	Material description	Typical density (Mg m^{-3})
	Individual cobble or gravel size particles	2.5
	Loose gravel on laboratory bench	2.0
	Loose sand on laboratory bench	1.5
	Clay within the lump	2.0

Table 8.9 Terms for description of secondary fine soils in coarse soils

Secondary constituent by weight	Term
up to 5%	Slightly clayey OR slightly silty
5% to 20%	Clayey OR silty
20% to (35%?)	Very clayey OR very silty

'Very' terms are only to be used when mass behaviour is judged to be that of a coarse soil, that the soil does not stick together when wet and remould.

- Very clayey secondary component will make the hand dirty and the soil may feel sticky and starting to stick together when wet, but not yet with sufficient fines to allow it to be moulded. The inclusion of a small additional quantity of fines will enable the soil to stick together when wet and remould and it would therefore be described as a fine soil.

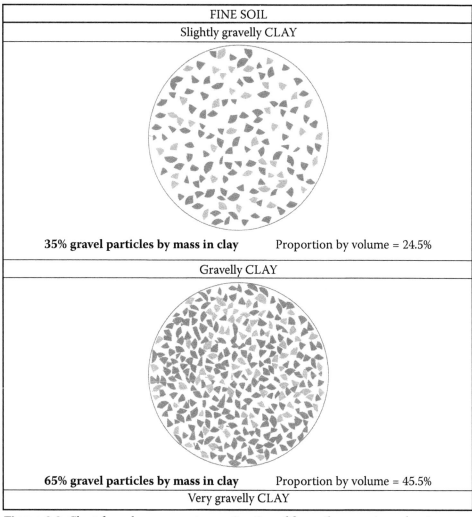

FINE SOIL	
Slightly gravelly CLAY	
35% gravel particles by mass in clay	Proportion by volume = 24.5%
Gravelly CLAY	
65% gravel particles by mass in clay	Proportion by volume = 45.5%
Very gravelly CLAY	

Figure 8.2 Chart for volume percentage estimation of fine soils, assuming a density adjustment ratio of 0.7.

EXAMPLE DESCRIPTIONS
- Clayey fine and medium SAND
  (5–20% fines behaving as clay, insignificant proportion of coarse sand)
- Very sandy silty GRAVEL
  (5–20% fines behaving as silt, 20–40% fine to coarse sand)
- Gravelly very silty SAND
  (20–40% fines behaving as silt, 5–20% gravel)

Although the adjectives silty or clayey are allowed to be used in fine soil mixtures, these terms should remain mutually exclusive when the fine soil is a secondary constituent. Although it should be possible to make a distinction between 'silty SAND' and 'clayey SAND', it is not realistically sensible to differentiate between, say, 'silty clayey SAND' and 'clayey SAND'. This exclusivity is not a requirement of the description standards, but represents a practical limit of logging possibilities.

The distinction between whether the fine soil fraction in this case is clay or is silt is not straightforward. The hand tests described in Chapter 4.8 can be used, but it is often difficult to separate out enough material to carry out the tests. In practice, a general feel for the soil grading can be quickly obtained by trying one or two of the hand tests such as feel and behaviour in air or water and observing the behaviour of the fines fraction on the hands.

The word order to be used when describing more than one secondary constituent is discussed in Section 8.1.6.

8.1.5 Secondary fractions in fine soils

The description of secondary constituents in fine soils is both simpler and more difficult. It is simpler as there are fewer options to consider, but more difficult as it is more subjective because the description is based on the feel of the soil and the hand tests rather than an assessment of the particle size grading. It should be remembered that mixed fine soils are limited to the series 'CLAY'–'silty CLAY'–'clayey SILT'–'SILT'. No other terms are allowed and the distinction within this series is made using the hand tests described in Chapter 4.8.

However, when there are secondary coarse soil constituents in a fine soil, these should be described and the terms for this are given in Table 8.10. It is noticeable that the percentages are very different from those used in coarse soils. This reflects the more dominant effect of the fine soil constituents so that more of the coarse secondary needs to be present in a fine soil to have the same effect as a markedly smaller proportion of secondary fine soil in a coarse soil. As with all soil mixtures, the proportion of each secondary constituent is assessed individually.

It is a clear outcome from this table that it is not possible to have 'very sandy very gravelly CLAY' or even 'sandy very gravelly CLAY' as the proportion exceeds 100% even before the CLAY is considered. It is theoretically possible to have 'sandy gravelly CLAY', although even this will not be common. The minimum values in the range of 35–65% can only leave room for 30% clay which may not be enough to make the soil behave as a fine soil; the increase of either or both of the coarse fractions to slightly more than 35% would render this mixture a coarse soil in that it would no longer stick together and remould when wet. Similarly, a 'slightly gravelly very sandy CLAY' is possible but not very likely, whilst a 'slightly gravelly slightly sandy CLAY' is possible.

In coarse soils the lower percentages in the definitions of the secondary constituent terms mean that this same problem does not occur.

Table 8.10 Terms for description of coarse secondary fractions in fine soil

Percentage of secondary coarse constituent	Term
up to 35%	Slightly sandy AND/OR slightly gravelly
35 to 65% NOTE sandy and gravelly is uncommon, but is possible	Sandy AND/OR gravelly
greater than 65%	Very sandy* OR very gravelly* *or describe as coarse soil

8.1.6 Multiple secondary fractions

A soil can have more than one secondary size fraction, but more than two is unlikely. This arises from the exclusivity of the silt and clay fractions when they are present as secondary fractions and the percentages that apply to the secondary constituent terms.

Combining the tables above for single secondary constituents into a comprehensive table which accommodates any number of secondary constituents gives the terms as in Table 8.11. As noted above, the proportion of each secondary size fraction is assessed separately and the total soil components in the coarse and fine fractions should not exceed 100%. This is the reason for the variation in the use of 'and', or 'and/or' in the tables throughout this chapter.

The distribution of soil mixtures can be represented within a tetrahedral space introduced in Figure 4.2 where the apices of the tetrahedron are clay–silt–sand–gravel. The very coarse fraction of cobbles and boulders plots outside this space.

Taking a coarse soil slice through this tetrahedron as shown in Figure 8.3 such that one side of the triangle is sand–gravel and the other side is fine soil, the distribution of the various fractions is shown in Figure 8.4.

The proportions of 5% and 20% given in Table 8.11 can be seen on the base line indicated by solid lines. As more fine soil is added the 5% and 20% percentage lines are apparent, moving away from the base line towards the apex. The point at which the coarse soil will start to stick together when wet and thus needs to be described as a fine soil, is shown at somewhere around the 35% point. The lines in this area are shown wavy in order to represent the uncertainty about the percentage at which this occurs.

A fine soil slice through the tetrahedron as shown in Figure 8.5 has one side of the triangle as clay–silt and the other corner as coarse soil and this is presented in Figure 8.6. In this diagram the major dominating effect of clay is apparent. The

Table 8.11 Terms for description of mixtures of fine and coarse soil

Term	Principal	Secondary	
		Coarse soil	**Fine soil**
Slightly clayey OR slightly silty AND/OR slightly sandy OR slightly gravelly	SAND		<5%
Clayey OR silty AND/OR sandy OR gravelly	AND/OR		5–20%
Very clayey OR very silty AND/OR very sandy OR very gravelly	GRAVEL	>20% or may stick together and so should be described as fine soil	
Very sandy OR very gravelly	SILT	>65% or may not stick together and so should be described as coarse soil	
sandy OR gravelly (AND only in theory)	OR	35–65%	
slightly sandy AND/OR slightly gravelly	CLAY	<35%	

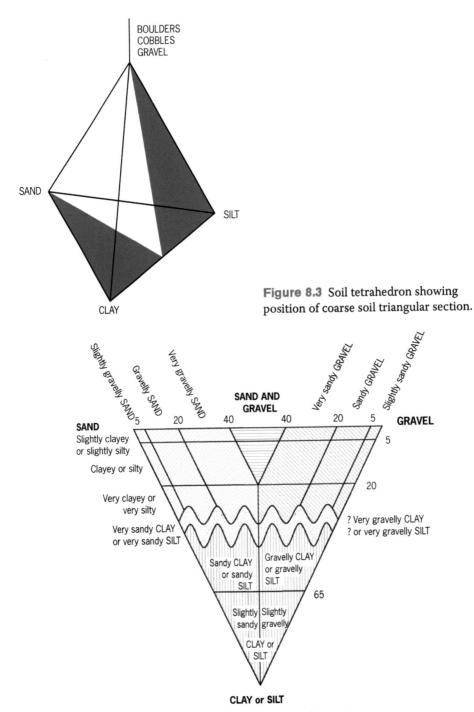

Figure 8.3 Soil tetrahedron showing position of coarse soil triangular section.

Figure 8.4 Coarse soil triangular section.

fine soil edge has no percentages given as this is the area assessed by hand tests. As coarse particles are added, these are included in the description using the terms in Table 8.11. At some point around the 65% coarse soil fraction the soil will stop sticking together and should be described as a coarse soil. Most of the lines on this

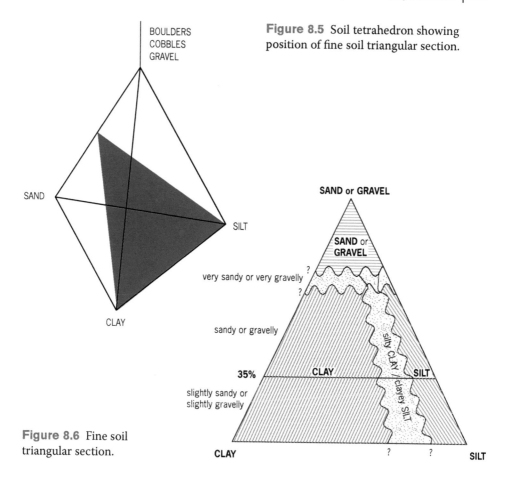

Figure 8.5 Soil tetrahedron showing position of fine soil triangular section.

Figure 8.6 Fine soil triangular section.

diagram are wavy as the assessment is made on plasticity and hand tests, not on the results of grading analyses.

The order in which the descriptors for multiple secondary constituents should be placed in a description is not dealt with satisfactorily in the Standards and is a matter that is open to debate. BS 5930: 1999 indicates that the dominant secondary fraction is given immediately preceding the principal soil name but this wording can be ambiguous.

Dominance can be taken to be:

- proportion when the secondary soil fractions are both coarse; this is straightforward and measurable and does not require judgement from loggers;
- proportion but accommodating the greater influence of fine soil constituents even in small quantities when the secondary soil fractions are both fine and coarse. This is difficult because it requires judgement from the inexperienced logger and it is a variable rule. Would the best descriptive order be 'sandy slightly clayey GRAVEL' or 'slightly clayey sandy GRAVEL'? There is an element of personal preference here, but it is not helpful to leave such matters open for decision in the field. Application of this sort of rule would depend on an accurate assessment of the relative proportions in the field and all experienced loggers would have to accept that this is often difficult.

A simplification of this variable rule would be to always put the fine after the coarse and immediately before the principal soil type even if the fine fraction does not dominate;

- grain size with either the coarser particles or the finer particles being taken to be more important.

The current Standards do not resolve this debate:

- BS 5930 Cl 41.4.4.5 'recommends that the dominant secondary fraction comes immediately before the principal soil term'.
- BS 5930 Cl 41.4.4.1 says that 'All soils should be described in terms of their likely engineering behaviour'.
- EN ISO 14688-1 Cl 4.3.3 states that 'adjectives describing secondary fractions shall be placed in the order of their relevance'.

Taking dominance, relevance and behaviour in combination gives rise to a sensible rule on word order which is:

- Increasing percentage when there are two coarse secondary constituent fractions, for example:
  - slightly sandy gravelly CLAY
  - slightly gravelly very sandy SILT
- Coarse and then fine if there are both fine and coarse secondary constituents present, for example:
  - very sandy silty CLAY
  - sandy clayey GRAVEL
  - very sandy slightly clayey GRAVEL
  - gravelly slightly silty SAND.

The description of secondary fractions should be checked against laboratory test results. This is particularly relevant for coarse secondary size fractions. The determination of particle sizes by sieving above 0.063 mm is sufficiently accurate to rely on the results of the laboratory analysis in preference to the visual observations of the logger, see Section 8.4. The accuracy of the particle size analysis methods used in determining the particle size distribution of fine soils cannot be relied upon in the same way. This means that care is required before amending a field description when applied to secondary fine constituents. Care is also required if the secondary particles are very coarse given that the size of sample must be large enough to be considered representative; large enough samples are the exception rather than the rule. The matter of checking descriptions against test results is discussed in Chapter 16.9.

8.2 Tertiary fractions

The description of tertiary constituents is less critical in the overall soil description as this aspect relates to the identification of origin of the deposit only and not to any aspect of its engineering properties. Therefore, a lesser degree of precision can be accepted in assessing the proportions present as they should only affect the geological interpretation, although accuracy is still, of course, required in identifying the nature of the tertiary constituent.

Table 8.12 Terms for description of tertiary soil constituents

Minor constituents before principal		
Adjectives (not quantified)	**Example terms**	**Recognition**
slightly	micaceous	shiny flecks
–	glauconitic	green or black specks
very	calcareous	reaction to acid (see below)

Minor constituents after principal		
with abundant	shell fragments	* including dimensions
with numerous	pockets of peat*	where size is otherwise not
with frequent	gravel size fragments of brick	defined is helpful in most
with occasional	flint gravel	cases
with rare	roots*	
	plastic bags	

The terminology for the proportion of tertiary constituents is defined only in qualitative terms and is given in Table 8.12. It is important to use these relative terms in a consistent manner. Although no quantification is included within the adjectives, they can be standardised on a site by use of reference samples.

The description of tertiary constituents can be included either before or after the principal soil type within the main description. This decision is to be made solely on the basis of making the description easy to read. Thus, for example, describing a clay as 'a shelly CLAY' might be easier than as a 'CLAY with frequent shells'. However, if any additional detail on the minor constituent is to be provided, the latter approach allows more flexibility, thus 'CLAY with abundant broken oyster shells'.

It may be that a tertiary constituent such as shells becomes sufficiently abundant to affect the soil properties; this raises it to a secondary constituent which is described as given in Chapter 8.1.

The use of dilute hydrochloric acid should be routine in field logging as a matter of discipline and good practice. The carbonate content can be significant because:

- carbonates can affect the engineering properties of a soil as a cementing agent;
- use of acid can assist identification of dolomitised rocks in carbonate terrains; and
- identification of the carbonate content can help stratigraphic identification.

The accurate description of the presence of carbonates in the soil or rock is good practice, but one that has progressively lapsed in recent years. The introduction of defined terms for the description of carbonate content in EN ISO 14689-1 requires that the practice be used routinely. This is good practice and should be followed.

The terms defined in EN ISO 14689-1 are as follows:

Apply dilute (10%) hydrochloric acid from a dropper and observe the reaction:

- No effervescence visible   Carbonate free
- Effervescence clear but not sustained   Calcareous
- Effervescence strong and sustained   Highly calcareous

TIP ON CARBONATE CONTENT TESTING
- In order to save a lot of effort, it is suggested that it is normally only necessary to report positive outcomes on a field log, i.e. the term 'carbonate free' does not need to be used in all descriptions such as when carrying out description of non-carbonate soils or rocks.
- If however, the formation is variably calcareous, identification of the carbonate free horizons can help make decisions on the stratification.
- The fact that testing by applying acid has been carried out and that the result is negative should be recorded on the sample or pit log description sheets.

An alternative classification by Burnett and Epps (1979) has found wide application in distinguishing between calcite and dolomite rich calcareous rocks of the Middle East:

- Strong reaction   Vigorous, frothy, audible effervescence
- Moderate reaction   Brisk, quiet, effervescence, constant stream of bubbles
- Weak reaction   Mild, faintly audible effervescence, small bubbles
- Negligible reaction   Negligible, inaudible effervescence, few bubbles

8.3 Description of widely graded soils

Soils in which most particle size fractions are present pose particular problems in soil description. First, there is coping with the description of the very coarse particles which are set in a matrix of finer soil particles, with all the problems outlined above. Second, there is the difficulty of separating out the coarse and fine fractions. This separation process is complicated because the sand tends to behave as part of the matrix, which is within the fine fraction and thus the normal split between coarse and fine soils is altered in terms of the assessment made in the field.

Norbury *et al.* (1986) proposed the matrix approach as suitable in these soils. This special process comprises separate description of the matrix and clasts where the clasts are those particles of gravel size or larger rather than the coarse and fine fractions. The description then considers the clay, silt and sand fractions as the matrix. These two constituents can be linked by normal adjectives to quantify whichever is the secondary constituent. These will combine to form a 100% soil before the very coarse particles are added back in. In practice the description of such soils will summarise the overall properties of the soil, which will be either clast supported or matrix supported. In the former, the clasts are in contact, and the soils behave in the manner of a coarse material and in the latter the matrix separates the clasts so the soil behaves in the manner of a fine soil.

The boundary between fine and coarse soil behaviour in these cases is not always straightforward, as it can depend on whether the relevant aspect of engineering behaviour is compressibility, shear strength or permeability. The proportions of clasts and matrix that affect these properties are different in each case. For example, the permeability is significantly affected by proportions of matrix or fines which are much lower than are needed to have an impact on the shear strength.

This descriptive practice for the explicit separation of the matrix has fallen into disuse since 1999. When in use this approach used the then current terms of 'with a little', 'with some' or 'with much' which were analogous to 'slightly', '−' or 'very'. Thus a description could have been:

- angular GRAVEL with some matrix of stiff grey sandy CLAY and with occasional cobbles;
- stiff brown sandy CLAY with much gravel and some cobbles of subrounded sandstone;
- firm grey clayey SAND with much rounded GRAVEL.

The adjectives used to describe the various soil proportions have since changed. This approach does not readily fit with the more rigid current separation into coarse and fine fractions and the use of adjectives only before the secondary term as detailed in BS 5930: 1999.

The inclusion of sand within the matrix remains the case in the description of concrete and grout where the sand is inseparable from the cementitious paste for the purposes of description.

The current approach for the descriptions of these three example materials would now be as follows:

- grey slightly sandy clayey GRAVEL with low cobble content;
- stiff brown slightly sandy gravelly CLAY with medium cobble content. Gravel and cobbles are subrounded sandstone;
- firm grey clayey very gravelly SAND. Gravel is rounded.

This differentiation between clast-supported and matrix-supported materials keeps cropping up in soil and rock description. The range from pure clasts to pure matrix provides a spectrum of behavioural characteristics which need to be described. The examples given above show the description of widely graded soils which are often glacial in origin. A similar situation occurs in the description of chalk where a distinction is made in structureless chalks between clast-supported Class Dc and matrix-supported Class Dm (Lord *et al.*, 2002). This gradation is also implicit within Approach 3 of the Rock Weathering Working Party (Anon, 1995) as indicated in Table 8.13.

Table 8.13 Supporting elements within a weathered profile

Likely behaviour	Zone	Proportions of material grades according to Approach 2 (Anon, 1995)
Matrix supported	6	100% G IV to VI (not necessarily all residual soil)
Matrix supported but with clasts affecting shear strength and compressibility	5	<30% G I–III >70% G IV–VI
Transitional between matrix supported and clast supported, partly dependent on whether considering shear strength, compressibility or permeability	4	30–50% G I–III 50–70% G IV–VI
Clast supported but with matrix controlling shear strength, compressibility and permeability	3	50–90% G I–III 10–50% G IV–VI
Clast supported but with matrix affecting permeability	2	>90% G I–III <10% G IV–VI
Clast supported	1	100% G I–III (not necessarily all fresh rock)

8.4 Description and classification of particle size grading

Description of a soil as a GRAVEL or as a SAND denotes equal proportions of coarse, medium and fine fractions as defined by the standards. This terminology is often used by lazy default rather than reflecting actual observations made. If equal proportions are not present, use of wording such as fine and medium indicates that these fractions predominate. This does not mean that there is no coarse sand present, just that the logger considers the proportion to be insignificant.

Examples of terms used to describe the grading of soils are given in Table 8.14, these examples are all in the range of sand for illustration but the sense of these terms can be used for any soil grading. The requirement of such terms is to try and use the minimum number of words while still conveying the complete and correct picture to the reader. It is also a matter of providing information on the overall character of the soil, not including too many categories just because there are one or two finer or coarser grains.

There are a number of terms used to classify the shape of a grading curve. The problem with these terms is that geologists and engineers have directly opposite definitions for these terms. To a geologist, a sediment that is well graded has a small range of grain sizes, whereas the well graded engineer's soil follows the Fuller curve with a wide range of grain sizes and proportions of the fractions to provide maximum density.

This is not a matter that should cause undue concern as the definition being used should normally be clear from the context when using the terminology provided in Table 8.14.

There are coefficients available which are based on the ratio of the particle sizes at 10% or 60% weight passing. The uniformity coefficient and coefficient of curvature classes are given in Table 8.15 from EN ISO 14688-2.

Table 8.14 Terms for description of grading

Grading terminology	Definition
Fine to coarse SAND OR fine, medium, coarse SAND	Grain sizes range from about 0.06–2 mm in roughly equal proportions. Often abbreviated as SAND.
SAND	Implies the full range of sizes present, often used as default but without the observation actually being made.
Fine and coarse SAND	Grain sizes of 0.06–0.2 mm and 0.6–2 mm predominate, i.e. the soil is gap graded.
Fine and medium SAND OR fine to medium SAND	Grain sizes between about 0.06–0.6 mm predominate, with only a small proportion of coarse sand (0.6–2 mm).
Fine SAND	Grain sizes of 0.06–0.2 mm predominate.

Table 8.15 Terms for classification of grading curves

Shape of grading curve	Uniformity coefficient $C_u = d60/d10$	Coefficient of curvature $C_c = (d30)^2/(d10 \times d60)$
Multi graded	>15	1 to 3
Medium graded	6–15	<1
Evenly graded	<6	<1
Gap graded	Usually high	Any (usually >0.5)

The value of these terms and the associated definitions is unclear. Of the grading curves shown in Figure 8.7 only one can be classified by these criteria, Curve 3 being multi graded. Even Curve 6, which is clearly gap graded, falls outside the relevant definition in Table 8.15 with a C_u of 4.3 and a C_c of 0.7. These terms will not appear on field descriptions and are of limited use even in report texts. The description of the shape of grading curves individually or as envelopes of several analyses is usually best made using diagrams such as Figure 8.7. The application of various combinations of secondary constituent terms is illustrated on some typical soil particle size grading curves in Figure 8.7.

The grading curves shown in Figure 8.7 would probably have the following descriptions:

1. CLAY: a typical clay soil such as London Clay with only fine constituents;
2. Slightly fine sandy CLAY: an estuarine fine soil which will behave as clay even though the bulk of the soil by mass is silt;
3. Slightly gravelly slightly sandy CLAY: a widely graded soil such as a glacial till, see Section 8.3;
4. Fine and medium sandy SILT: a mixture typical of many alluvial soils;
5. Fine sandy SILT or silty fine SAND: a soil on the borderline between fine and coarse, which might or might not stick together;
6. Very gravelly clayey SAND: a gap graded soil lacking the medium and coarse sand fractions that might have been expected;
7. Very gravelly SAND: a typical river terrace deposit in which the proportion of fines can be critical;
8. Sandy GRAVEL: an outwash or wadi deposit that is well graded and close to the ideal Fuller curve (which is the grading for the soil to have a maximum density of packing).

8.5 Other information

Any other information or observations on the soil or rock material should be included at this stage of the description process. Items to be incorporated could include:

- abnormal mineralogy;
- presence of vugs or voids, particularly if interconnected such as shown in Figure 8.8;
- strange colours;
- odours.

Sample descriptions and field logs should also include any other information available to the logger about the origin of the soil or effects of drilling and sampling on the sample. This will include all information not yet incorporated into the description. Information falling into this category can come from observations made by the logger, the driller (written or verbal), the miner, the driver, or anyone else involved with the site work. Information presented under this category will comprise further and non-standard details which enhance the description already made. It can therefore include information on the formation of the borehole or trial pit, on water strikes, on flush returns, on stability or any matter relevant to the assessment of the ground in situ.

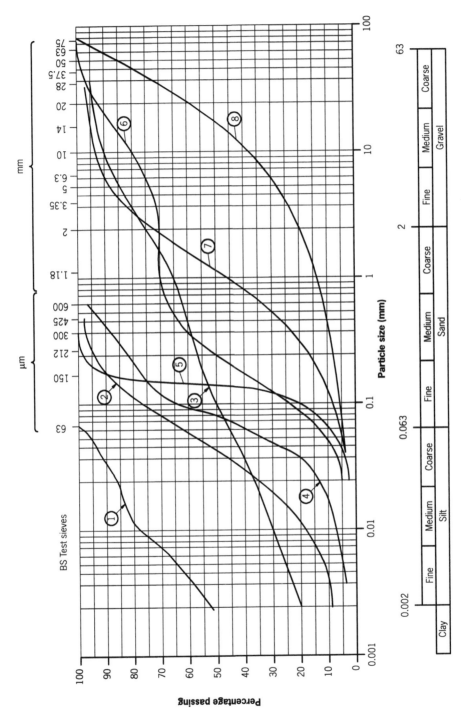

Figure 8.7 Illustrative grading curves and descriptions.

Figure 8.8 Example of other information in the form of voids in the core.

In particular any observations that the recovered sample may be unrepresentative of the ground should be included. The reasons for this situation arising can include:

- disruption – during and after sampling arising from physical disturbance;
- degradation – mechanical (e.g. frost damage) or chemical (e.g. crystal growth associated with pyrite degradation) during storage;
- softening (increase in water content) affecting apparent consistency, e.g. during sampling, from chiselling effort, moisture migration from sand inclusions during storage;
- desiccation (decrease in water content) giving increase in consistency, e.g. from unwaxed sample, long storage;
- lack of representativeness – loss of fines in the water when shell emptied (whether there is 3% or 6% silt in a sand is important in dewatering or grouting);
- lack of representativeness – very coarse soils shown by angular broken particles;
- lack of representativeness – sample too small;
- contamination – presence of oils, smells, odd colours and sharp items which are potentially hazardous. It is rather more than just a courtesy to professional colleagues to remember to inform the laboratory if sharp or hazardous objects are found in a sample bag. Just because the logger found such inclusions by surprise is no reason to wish a similar shock to others;
- foreign bodies – items introduced by the sampling process e.g. sandstone gravel from the river deposit carried down and included within the tube sample from the slurry at the base of the hole, brick fragments caving into the hole from made ground, or green grass or clinker fragments from the sample being tipped out before being bagged.

9 Geological Formation

The description of the soil or rock material is completed by the formation name. The formation name is taken from the geological map of the site area and some examples are given in Table 9.1. The naming of the geological formation does not replace any aspect of the description as given throughout this book. The formation name completes the description by dividing the logged sequence into geological units or formations and is carried out by the logger who is the person who has actually inspected all the samples and exposures that are reported on the field logs.

Two points arise from these examples. The first is that an indication of levels of uncertainty is included by the use of the terms possible, probably or question marks. This is good practice and often arises at the vertical or horizontal boundaries of the formation.

The second point of note is the inclusion of weathering classifiers in the formation. As noted in Chapter 10, the description of the weathering features is an important element within the main description, but the classification needs to be separated out and hence inclusion as part of the formation name is a sensible option.

Table 9.1 Example formation names

Some typical formation names taken from survey maps:

- ALLUVIUM
- ALLUVIAL CLAY
- ESTUARINE DEPOSITS
- RIVER GRAVEL
- TERRACE GRAVELS
- FLUVIO-GLACIAL DEPOSITS
- Probably BLOWN SAND
- Possibly GLACIAL SAND AND GRAVEL
- MADE GROUND/RIVER GRAVEL?
- MADE GROUND/BOULDER CLAY?
- Completely weathered SHERWOOD SANDSTONE
- LONDON CLAY
- Probably REWORKED LONDON CLAY
- Completely weathered MERCIA MUDSTONE
- COAL MEASURES which with finer resolution could be named as
  - Weathered WICKERSLEY ROCK or
  - Residual MUDSTONE
- SHERWOOD SANDSTONE FORMATION
- FLAMBOROUGH CHALK
- Distinctly weathered MERCIA MUDSTONE

In theory the application of formation names is straightforward but this is not always the case. For example, many formation names have been changed over the years as the national survey falls into line with revised international stratigraphic conventions. So, for instance in the UK, Keuper Marl has been renamed as Mercia Mudstone and Upper Chalk has become the White Chalk Group. Guidance on old and new formation names used in British stratigraphical nomenclature is available from CIRIA SP 149 (Powell, 1998). Unfortunately it takes many years for the survey publications and maps to catch up with such changes so some extrapolation may be necessary. The safest course is to use the formation name on the published map, even if this is known to have been replaced. Although some boldness in applying a new name can be perfectly reasonable, this can also result in problems. A practical and robust view needs to be taken in this matter, as outlined in Table 9.2.

There are those who believe that to include the geological formation on a field log is a step too far into classification and interpretation. This is not realistically the case as the logger is the best person to make the assessment of the formation boundaries being the only person that has seen all the samples, core or exposure.

There is a caveat to this view in those cases where naming the formation is unduly sensitive for one reason or another. These reasons usually occur in sequences where naming of the formation boundary is difficult, because of variability in the sequence or where there is a need for specific and local knowledge. The instances where specialist knowledge may be invoked include the naming of marker fossils, the identification of associations of depositional features, to the naming of individual coal or other mineral seams.

However, the most common case where formation names are omitted relates to investigations in legal cases or claims. In these instances, it is permissible to omit a formation name. Nevertheless, the default and normal situation should be for the formation name as indicated in published geological information to be included and some typical examples are given in Table 9.1.

Logging of geological successions should include all pertinent information on marker beds where these are stratigraphically significant. Markers can include particular beds or fossils, or other geological associations of features, such as ash bands,

Table 9.2 Naming a geological formation

Status of map	Procedure	Example
Available map has new formation names.	Use new formation name.	Mercia Mudstone Group
Available map has old formation name, but new formation name is applicable using common sense.	Use new formation name with explanation in report text as to protocol followed.	London Clay Formation
Available map has old formation name, but new formation name is not reliably applicable.	Use old formation name.	Middle Chalk
Available map has old formation name, but new formation name is not readily applicable.	Omit formation name and explain why in text.	—

TIP ON NAMING OF STRATIGRAPHIC MARKERS

- The presence of all sequence markers should be recorded as a precursor to naming. Given the lack of local or formation-specific knowledge held by most engineering geologists, this naming is usually carried out by geologists who have the relevant specialised knowledge on the formation, or the basin, or the coalfield and who are better placed to do so.
- The initial identification in the engineering geological field log will be at a fairly high level. The marls or ash bands will be recorded as present with no additional information. The fossil identification may only be at the genus level, but that can still be useful.
- Naming of fossils to species level will usually require the input of special expertise.
- Similarly it would not be realistic to expect most engineering geologists to name individual markers such as marls, ash bands or coal seams, for example. That would sensibly require input from the local regional or pit geologist.
- Where specialist input is provided by others, the naming can be included on field logs with due attribution.

weathering effects or facies changes. It is good practice for the logging geologist to make themselves aware of the lithostratigraphic succession and thus which if any of these features may be diagnostic and should therefore be recorded. The logger needs to be aware of the expected succession if they are to prepare an accurate and complete log (see Chapter 16). The incorporation of relevant information on the stratigraphic succession is only really likely to be achieved if the geological model is understood. In order to obtain this background knowledge the geologist may need to read up the relevant references about the formation that is being investigated, or attend some specific briefing or training before going to site.

However, most engineering geologists are mobile geographically (and therefore geologically) and are therefore not often in the position of having sufficient knowledge to name these bands or fossils without conjecture. Such conjecture is of course to be avoided.

10 Weathering

10.1 Weathering of soils

The impact of weathering on soils is generally underestimated. In northern Europe, it is often assumed that the weathered profiles were removed by ice action. However this is not always the case and extensive and significant weathering profiles can be found, these being due either to fossil weathering or to more recent quaternary weathering associated with differing climatic regimes that have affected the ground. Many of the mass slope movements that are present over the region owe their initiation to processes that started under periglacial conditions and the failures at Walton's Wood (Early and Skempton, 1974) and Carsington (Skempton *et al.*, 1991) are just two examples drawn from a wide literature on the subject showing how these relict features can have an impact on the construction and safety of engineering projects. In other parts of the world which were not subjected to the direct effects of ice, deep weathering profiles are present which require special consideration.

Under any particular environmental conditions, there are processes taking place which have a marked effect on the engineering characteristics of the soil. Recognition of the effects of these processes and their impact on the ground profile is fundamental to the correct assessment of the engineering properties of the soil mass. If the weathering profile is not correctly identified, lack of understanding can lead to inappropriate engineering designs. For instance, the presence of low angle shears in periglacially affected soil is likely and these will significantly reduce the strength of the soil.

Weathering subjects the ground to a series of physical and chemical processes. These include stress fluctuations, volume changes, oxidation and reduction which result in:

- reduced strength   softening
- loss of structure   destructuring
- changes of colour   mottling
- leaching of minerals   removal of sulphides
- growth of new minerals   gypsum growth
- reduced confining pressure   more fractures.

The result of these processes on a typical mudrock (London Clay for example) will include the features summarised in Table 10.1.

The fundamental requirement in describing any weathered profile is actually to describe what is seen. A logger working through a weathered profile displaying these features may or may not recognise what they are seeing, but will split the profile into a number of strata, which are usually thin units with boundaries

Table 10.1 Summary of weathering processes acting on a mudstone/overconsolidated clay

Feature	Unweathered condition	Weathered condition
Colour	Blue grey	Brown/orange
Fissure spacing	Very close	Extremely close
Bedding	Present	Absent
Structure	Rectilinear	Lensoidal

selected on the basis of these changes. Thus, on completion the alert logger will have described the weathering profile.

In order to provide background, the individual physical and chemical processes that are evidenced in these weathered profiles are outlined below in Table 10.2 after Spink and Norbury (1993).

The weathering processes and their visible manifestations outlined in Table 10.2 are those typically applicable to stiff overconsolidated soils (e.g. argillaceous deposits from any age but most commonly the Mesozoic and Quaternary clays and also glacial tills) which may well include completely weathered rocks (Spink and Norbury, 1993). The thickness of this weathered profile is generally limited and is not often more than 5 m or maybe 10 m, but this is the layer in which a high proportion of engineering works are carried out and so these weathered profiles are very significant.

Examples of the description of weathered soils are given below. These descriptions use all the normal descriptive terminology to record the features present. The

EXAMPLE DESCRIPTIONS

- Firm to stiff lensoidally fissured and cracked brown mottled light grey CLAY. Lensoidal fissuring is extremely closely spaced (3/5/10). Contraction cracks are subvertical medium to closely spaced rough light grey gleyed. Occasional gypsum crystals (destructured LONDON CLAY)
- Stiff fissured brown mottled grey slightly sandy CLAY with occasional rounded gravel size lithorelicts of very weak mudstone. Fissures are randomly orientated very closely spaced (weathered CARBONIFEROUS MUDSTONE).

BACKGROUND ON GLEYING

- Gleying is the process of reduction in waterlogged soils, such as occur in very cold and permafrost conditions. In the waterlogged soils, water replaces air in the pores and so the available oxygen is quickly used up by microbes feeding on organic matter in the soil. This creates anaerobic conditions and leads to the reduction of iron from its oxidised (ferric) Fe^{3+} state to its reduced (ferrous) Fe^{2+} state. The reduced compounds are more soluble in the soil water and the removal of the iron leaves the soil a grey or light grey colour.
- Examples of gleyed fabric are shown in Figure 10.1.

Table 10.2 Physical and chemical weathering processes and features to be described

Weathering class	Physical weathering processes	Observable effects	Bedding	Chemical weathering processes	Observable effects
Reworked		• Occasional randomly orientated lithorelicts • Occasional included foreign material	No original bedding		• Brown and light grey mottled clay • Gypsum crystals rare or absent.
Destructured		• Frequent randomly orientated lithorelicts • Numerous horizontally aligned lithorelicts • Lensoidally fissured	Bedding increasingly disturbed	Increasing leaching Increasing reduction (gleying)	• Brown and light grey mottled clay • Centre of relicts and fissure blocks brown • Gypsum crystals common
Weathered			Bedding undisturbed	Oxidation complete	• Brown clay with light grey gleying on fissures • Gypsum crystals common • Brown clay • Gypsum crystals common
Partially weathered		• Fissure spacing decreases • Increasing likelihood of occasional slightly polished fissures	Bedding undisturbed	Increasing oxidation	• Brown clay around fissures • Centre of fissure blocks grey • Occasional gypsum crystals • Grey clay with brown staining • Rare gypsum crystals on fissures
Unweathered		• Original fissured clay	Bedding undisturbed		• Grey clay

Physical weathering processes (ranging across classes):

- Solifluction shears
- Increasing likelihood of polished shear surfaces
- Periodic moisture content and volume changes
- Moisture content and volume changes associated with stress relief, periglacial and chemical processes
- Random accommodation shearing by solifluction
- Fabric disturbance from chemical weathering
- Subhorizontal shearing by solifluction
- High angle shearing by periglacial heave/collapse
- Deep shearing by periglacial processes
- Modern desiccation cracks
- Old subvertical contraction cracks
- Fissure formation by ice lensing
- Fissure formation by stress relief

(a)

(c)

(b)

Figure 10.1 Examples of gleyed fabric: (a) shows gleying following vertical fissure surfaces (possible desiccation cracks) and silt and find sand partings, (b) and (c) show the gleying following root tracks (some root material still present) as well as short fissures and shear surfaces.

classification of these materials, where a suitable classification is available, is discussed further below.

10.2 Rock weathering

The historical background to the description, or rather more accurately the classification, of weathering has been discussed in Chapter 2.4. It has been noted that prior approaches to the description of weathering have failed because classifications applied directly in the field have replaced complete description. The logging geologist tended to 'describe' a rock as moderately weathered, highly decomposed or whatever, at the expense of accurate and detailed field description. The inappropriateness of this approach is shown by the range of weathering processes illustrated in Figure 10.2 and the implausibility of trying to cover all these options within a single, all encompassing weathering classification is apparent.

The approach put forward in BS 5930: 1981 (as well as IAEG, 1981 and ISRM, 1981) was not generally applicable in the northern hemisphere as inappropriate weathering processes were incorporated so this approach was not used in the UK. In BS 5930: 1999 the recommendations of the Working Party on rock weathering (Anon, 1995) were adopted into UK practice. However, EN ISO 14689-1 stepped backwards and reverted to the earlier approach which had been discarded from practice in the UK and those other countries that base their practice on BS 5930. The reasons for this being a backward step are outlined by Hencher (2008).

Current UK practice is to continue with the principles laid out in Anon (1995). The descriptive approach together with four classification approaches outlined in Anon 1995 are summarised in Table 10.3. However, because of direct conflict

(a)

(c)

(b)

Figure 10.2 Exposures showing different styles of weathering: (a) shows weathering by disintegration with joints forming and opening but otherwise the rock material only being slightly affected, (b) shows granite in Hong Kong weathering by decomposition migrating inwards from the original joints and with fresh and residual material present in the same exposure, and (c) shows weathering by dissolution along joints in a Portuguese limestone leaving voids and fresh material with small quantities of insoluble residues on joint surfaces (photographs courtesy of Steve Hencher).

between Approach 2 and EN ISO 14689-1, where the same terms are used for mass weathering but with different definitions, Approach 2 has to fall from national standards and usage. The Approach 3 zones are defined on the basis of Approach 2 grades (see below) which are no longer available and so Approach 3 also has to fall by the wayside.

This is the position that has been written into BS 5930: 1999 as shown by the greyed areas in Table 10.3. Therefore, the position now is that a Eurocode compliant description cannot use Approaches 2 and 3. If, however, an investigation is prepared not to be fully compliant with EC7, there is no reason why the weathering profiles cannot be described and classified in accordance with Anon (1995), rather than having to slavishly follow EN ISO 14689-1. Note that an investigation that follows such an approach is perfectly permissible although it could not be subsequently claimed that the investigation and therefore the design were wholly EC7 compliant.

Full details of this scheme are given in Anon (1995) and the process for selection of the most appropriate approach to follow is presented in Figure 10.3.

Table 10.3 Approaches to weathering description and classification

Approach	Scheme	Basis	Applicable to	Examples
1	Description	Mandatory description of what is seen	All soils and rocks	
2	Grades I–VI	Discolouration and weakening	Uniform materials Medium strong or stronger	Gritstones Coal Measure sandstones
3	Zones 1–6	Proportions of Approach 2 grades	Heterogeneous masses Medium strong or stronger	Granites
4	Classes A–E	Fabric changes and weakening	Material and mass features Weak or weaker rocks	Mercia Mudstone Lias Clay Coal Measure mudstones
5	Special cases			Chalk Karst Arid climate weathering

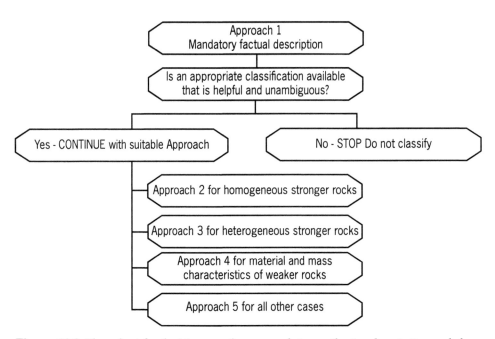

Figure 10.3 Flow chart for decisions on the approach to weathering description and classification to be adopted (Anon, 1995).

The fundamental requirement within this procedure is for accurate and complete description. This can be followed by classification but only if an applicable and useful scheme is available. Classifications are as diverse as the range of rock types together with the variety of weathering processes superimposed on that range. Formal classification on field logs is not mandatory and should only be applied where useful, well established, unambiguous systems are available. These classification systems can either be available from the literature and based on similar geological conditions elsewhere or compiled on a site specific basis. The use of existing schemes will provide benefit from allowing precedent experience from other sites to be incorporated into the assessment of the investigation results.

Where a site specific scheme is established, there may be little or no precedent experience to assist the design process. If this approach is being followed, the sort of features to be identified and included when preparing a classification scheme are included in the weathering features that need to be described and are incorporated into Approach 1 as summarised below.

10.3 Approach 1: description of weathering

It is fundamental that the effects of weathering are recorded in the description of soils and rocks for engineering purposes. The first and essential step in any weathering assessment is therefore to prepare a full factual description of the degree and extent of weathering and of the evidence for weathering having taken place together with a description of the character of any products that remain from the weathering process. In comparison with some of the previously adopted weathering 'classification' schemes, this does not require sight of the complete weathering profile, which would be a significant advantage in many engineering investigations which may not however go to sufficient depth for this to be realised. The description will form the basis of any classification of the weathering profile, as this can subsequently be applied on the basis of the information included in the description.

Description of the weathering features named as Approach 1 is considered mandatory as it forms an essential part of a full description. The combination of the defined terminology throughout this book together with non-technical terms allow the logger to emphasise those features resulting from weathering and the state of the processes that are visible in the exposure.

The information to be recorded within a description following Approach 1 includes the aspects listed below:

- The degree and extent of colour changes which are associated with weathering should be identified and described:
  - The degree of colour change can be reported using relative terms such as 'slightly discoloured' or 'highly stained' which can be enhanced by giving the colour of the staining;
  - The extent of colour change can use relative terms such as 'locally discoloured' or 'penetratively (5 mm) discoloured';
  - It is up to the logger to identify the original fresh colour if possible; it is then perfectly reasonable to presume that any colour changes are due to weathering. Comment on this aspect would normally only be required when this presumption is invalid, which would be something worth reporting.

- The original strength of the soil or rock and any changes in that strength associated with weathering should be identified and described:
  - The description of the strength would make use of the normal defined terms given in Chapter 5 and this can be enhanced with measurements such as 'strength reduced to very weak up to 50 mm from joints';
  - The description can also include non-defined terms in accordance with non-technical language such as 'weakened within weathered zones'.
- The fracture state at the various levels of the weathering profile should be described, with special attention being paid to any changes in the spacing and in the character of the fractures. For example, reference to Table 10.2 indicates that the fissure spacing reduces and becomes less ordered as weathering progresses:
  - The description should use the normal defined terms and measurements and indicate where changes in these spacings are thought to be due to weathering.
- The presence and nature of any weathering products should be described. These will normally be expected to be associated with infill or penetration from the fractures, but can occur elsewhere within the material:
  - Use appropriate soil or rock descriptive terms for products and quantify their extent.

An unusually detailed example of the description of weathering features is given below. It is not expected that this full level of detail will be available or recorded except in particular circumstances. The intention of this example is to illustrate how the approach can be applied and to show the combination of standard defined terms together with normal English terms. This approach has been found to be workable and to create accessible descriptions in which the reader can readily comprehend the weathering profile. On this basis alone, this Approach has to be considered a success and one which should be followed.

At this point the logger has to decide whether to classify or not. As noted above this is not mandatory and should only be carried out when an appropriate classification scheme is available or can be compiled. The classification approaches included in Anon (1995) are outlined below.

The classification schemes are split into those for rocks that are medium strong or stronger, in which case Approaches 2 and 3 should be followed, and those that are weak or weaker, in which case Approach 4 should be followed. There is also

---

EXAMPLE DESCRIPTION
- Strong thinly bedded light and dark grey medium grained SANDSTONE (CARBONIFEROUS SANDSTONE).
  Weathering:
    discontinuities heavily stained black;
    with light orange brown discolouration penetrating up to 30 mm;
    slight loss of strength on discontinuity walls up to 10 mm penetration;
    extremely closely spaced random fractures in 40 mm thick zones around joints;
    occasional sand infill up to 0.5 mm in joints.
  Joints are....(description of the joint sets follows, see Chapter 11).

within Approach 5 accommodation of weathering classification schemes for other formations or geological settings. Examples for chalk, karst and tropical weathering are given.

10.4 Approaches 2 and 3: classifications for homogeneous stronger rocks

The Approach 2 classification given in Table 10.4 is based on the progressive weathering of intact rock material. This approach is normally only applicable to rocks which are medium strong or stronger. The basis of the classification is the material strength as determined by simple tests that can be carried out in the field. The test results are subject to variation depending on the recovered condition so care needs to be exercised in using this approach; the use of index tests to support the field classification is recommended.

The classification within Approach 2 is generally and widely applicable. However the classification terms have been used in most other classification schemes so great care is needed when applying or reading these terms to understand which definition is in use. Approach 3 uses the grades from Approach 2 for the classification and is suggested for use in rocks which develop weathering profiles which combine the presence of relatively strong and relatively weak materials within the mass, much as envisaged by the weathering classifications put forward in BS 5930: 1981 and EN ISO 14689-1. It is not always the case that rocks weather in this way, nor that this classification is the best way to separate out zones of very different engineering character. It should be noted from the terms in Table 10.5 that even in Zone 1 the mass is not entirely unweathered material, as it includes up to Grade III material; similarly Zone 6 includes Grades IV to VI materials.

It can be seen from the Approach 3 terms given in Table 10.5 that there will be problems applying this in the field because boreholes are not usually sufficient for appropriate identification of the mass and that assessment of the volumes will be difficult. Before applying this approach it must be asked whether the use of a zonal or rock mass classification approach would actually be more relevant to the project and thus to the ground description. This is generally the case in practice so Approach 3 is not widely used. As noted above under the behaviour of certain soils, the concepts of progressive change of properties arising from progressive destructuring of a soil or rock mass can be helpful in other areas of description.

Table 10.4 Approach 2: classification for uniform materials

Grade	Classifier	Typical characteristics
VI	Residual soil	Soil, retains none of original texture or fabric
V	Completely weathered	Considerably weakened Slakes Original texture apparent
IV	Highly weathered	Does not readily slake when dry sample is immersed in water
III	Moderately weathered	Considerably weakened, penetrative discolouration
II	Slightly weathered	Slight discolouration, weakening
I	Fresh	Unchanged from original state

Table 10.5 Approach 3: classification for heterogeneous masses

Zone	Proportions of material grades	Typical characteristics
6	100% G IV to VI (not necessarily all residual soil)	May behave as soil.
5	<30% G I–III >70% G IV–VI	Weak grades control behaviour. Corestones may be significant.
4	30–50% G I–III 50–70% G IV–VI	Rock framework contributes to strength; weathering products (matrix) control stiffness and permeability.
3	50–90% G I–III 10–50% G IV–VI	Rock framework controls strength and stiffness; matrix controls permeability.
2	>90% G I–III <10% G IV–VI	Weak materials along discontinuities affect shear strength, stiffness, permeability.
1	100% G I–III (not necessarily all fresh rock)	Behaves as rock.

10.5 Approach 4: classification for heterogeneous weaker rocks

In those rocks in which the weathering develops and enhances the heterogeneity such that the material and mass characteristics are best combined into a single classification scheme, use of Approach 4 is suggested. The scheme is detailed in Table 10.6 and will generally be found to be most applicable in rocks that are weak or weaker. In practice, this approach is found to be most suitable for many sedimentary rocks and overconsolidated clays.

In practice this is the most commonly used of the Approaches in current practice. This is largely because a number of formation-specific classifications for overconsolidated clay/weak rock formations which have undergone a combination of the physical and chemical weathering processes outlined in Table 10.2 were already available in the literature. These were mapped on to the general Approach 4 classification (see tables below) by Spink and Norbury (1993). The formations that these can be applied to are specifically the London Clay, Mercia Mudstone, Lias Clay and Oxford Clay, but there is no reason why this Approach cannot be used for any similar material. One such example is the Coal Measures mudstones which are not covered explicitly here, but can be classified in accordance with Table 10.6. The formation-specific schemes outlined below illustrate the application to these units and therefore how these could be translated onto other units.

It has been noted that the classifications arising can appear with the geological formation. Examples of how this might be carried out are shown below.

EXAMPLES OF WEATHERING CLASSIFICATIONS
- Either partially weathered LONDON CLAY
- Or LONDON CLAY Class B

- Either destructured MERCIA MUDSTONE GROUP
- Or MERCIA MUDSTONE GROUP Class Db

Table 10.6 Approach 4: general classification for weak rocks

Class	Classifier	Typical characteristics
E	Reworked or residual	Matrix with occasional altered random or 'apparent' lithorelicts, bedding destroyed. Classed as reworked when foreign inclusions are present as a result of transportation
D	Destructured	Greatly weakened, mottled, ordered lithorelicts in matrix becoming weakened and disordered, bedding disturbed
C	Distinctly weathered	Further weakened, much closer fracture spacing, grey reduction colours
B	Partially weathered	Slightly reduced strength, slightly closer fracture spacing, weathering penetrating in from fractures, brown oxidation colours
A	Unweathered	Original strength, colour and fracture spacing

10.6 Material specific weathering schemes

A selection of the formation-specific schemes is presented below. The format of these tables is to provide the classification in accordance with Approach 4 as given in Table 10.6 in the first two columns. The example description in column 3 is that given in the original paper but updated to accord with current standard terminology. The zonation provided by the authors in the original papers is included in column 4 in Tables 10.7–10.10.

The weathering classification for London Clay based on Spink and Norbury (1993) and after Chandler and Apted (1988) is shown in Table 10.7.

The weathering classification for Lias Clay based on Spink and Norbury (1993) and after Chandler (1972) is shown in Table 10.8.

The weathering classification for Mercia Mudstone based on Spink and Norbury (1993) and after Chandler (1969) and Chandler and Davis (1973) is shown in Table 10.9. The descriptive terms used by Chandler are also included for reference (in italics) but these should no longer be used.

The weathering classification for Oxford Clay based on Spink and Norbury (1993) and after Russell and Parker (1979) is shown in Table 10.10.

10.7 Approach 5: special cases

There will always be certain geological situations where the weathering does not follow the patterns envisaged in the Approaches outlined above. This requires the classification of the weathering to be tailored to be appropriate to the rock and the superimposed weathering processes, making any such scheme specific to those conditions and unlikely to be applicable more widely. Some of the cases on which guidance is published are summarised below, but other weathering classifications can always be compiled along the lines suggested by these precedent systems. The examples cited below cover special lithological varieties (such as chalk), particular processes such as karst dissolution or particular forms of weathering, namely tropical.

10.7.1 Chalk

Chalk is a special case which crops out or underlies large areas of northern Europe and further afield. The area of the UK underlain by chalk incorporates a section

Table 10.7 Weathering classification for London Clay

Class (subdivisions after Spink and Norbury, 1993)		Example descriptions (Note – classes and example descriptions as BS 5930: 1999)	Zone (Chandler and Apted, 1988)
Reworked	E	Stiff becoming firm with depth sheared cracked brown and light grey mottled CLAY with rare up to 5 mm lithorelicts. Rare rounded flint gravel. Shears are randomly orientated very closely spaced discontinuous, slightly polished and subhorizontal closely to medium spaced of 0.5–1.5 m extent polished light grey gleyed. Desiccation cracks are subvertical closely spaced rough.	Head
Destructured	Db	Firm sheared cracked brown and light grey mottled CLAY with occasional up to 5 mm, randomly orientated, firm to stiff lithorelicts. Shears are inclined medium spaced up to 1.0 m extent polished light grey and gleyed. Desiccation cracks are subvertical closely spaced rough light grey gleyed. Occasional gypsum crystals.	IV
	Da	Firm cracked brown and light grey mottled CLAY with numerous up to 10 mm, horizontally aligned, firm to stiff lithorelicts. Desiccation cracks are subvertical closely spaced rough light grey gleyed. Occasional gypsum crystals.	IV
Distinctly weathered	Cc	Firm to stiff extremely closely lensoidally fissured and cracked brown mottled light grey CLAY. Desiccation cracks are subvertical medium to closely spaced rough light grey gleyed. Occasional gypsum crystals.	IIIb
	Cb	Stiff extremely closely fissured sheared brown occasionally mottled light grey CLAY with occasional partings of brown silt. Fissures have light grey gleying. Shears are rare slightly polished up to 0.1 m extent and rare subhorizontal, polished, light grey gleyed up to 1.0 m extent. Desiccation cracks are subvertical medium spaced rough light grey gleyed. Occasional gypsum crystals.	IIIa
	Ca	Stiff extremely to very closely fissured brown CLAY with occasional partings of brown silt. Fissures are rare slightly polished up to 0.1 m extent. Occasional gypsum crystals.	IIIa
Partially weathered	Bb	Stiff to very stiff very closely fissured brown and grey mottled CLAY with occasional partings of brown silt. Rare gypsum crystals on fissures.	IIb
	Ba	Very stiff very closely fissured grey occasionally mottled brown, CLAY with occasional partings of brown silt. Fissures have brown staining.	IIa
Unweathered	Ab	Very stiff very closely to closely fissured grey CLAY with occasional partings of grey silt.	I
	Aa	Very stiff closely fissured grey CLAY with occasional partings of grey silt.	I

Table 10.8 Weathering classification for Lias Clay

Class (subdivisions after Spink and Norbury, 1993)		Example descriptions (Note – classes are as BS 5930: 1999) *Colloquial descriptions from Chandler are for information.*	Zone (Chandler, 1972)
Reworked	E	Stiff light grey and light brown mottled, becoming grey with depth, CLAY. In upper parts frequent coarse sand size haematite and rotated lithorelicts up to 1 mm. In lower parts lithorelicts are rotated and up to coarse gravel size (30 mm). Frequent short polished shear surfaces throughout. *Mottled light grey/light brown, becoming grey with depth, CLAY. At shallow depth extensive oxidation and gleying, frequent haematite pellets, small rotated lithorelicts up to 1 mm, desiccation cracks. At depth rotated lithorelicts up to 30 mm. Minor shearing throughout.*	Landslip/ solifluction
Destructured	D	*Not observed – probably absent on slopes. Horizontally orientated lithorelicts anticipated.*	IV
Distinctly weathered	C	Very stiff to stiff fissured light brown CLAY. Fissures extremely to very closely spaced (10–30 mm) with remoulded light grey gleyed clay up to 20 mm thick and containing occasional polished surfaces. Lithorelicts horizontally aligned. Frequent gypsum crystals. *Pale brown fissured CLAY with light grey gleying on fissures. Fissures 10–30 mm spacing. Remoulded gleyed matrix along and around fissures occupies less than 50% in thin section and contains minor shearing. Lithorelicts (fissured blocks) have horizontal bedding. Common gypsum crystals.*	III
Partially weathered	Bb	Very stiff fissured and sheared bluish grey CLAY. Fissures generally very closely spaced stained brown with discolouration penetrating. Shears show 1–2 mm displacement. Infrequent gypsum crystals. *Grey/blue-grey CLAY. Fissures 20–100 mm spacing with brown staining in areas parallel to fissures. Minor shears with 1–2 mm displacement. Infrequent gypsum crystals.*	IIb
	Ba	Extremely weak to very stiff fissured bluish grey MUDSTONE/CLAY. Fissures generally very closely spaced stained brown. Rare gypsum crystals. *'Soft' blue grey MUDSTONE/CLAY. Fissures 20–100 mm spacing, brown stained. Rare gypsum crystals.*	IIa
Unweathered	A	Extremely weak to very weak fissured bluish grey MUDSTONE. Fissures generally closely spaced. Bedding horizontal. *'Soft' blue-grey MUDSTONE. Fissures greater than 100 mm spacing. Bedding horizontal.*	I

Table 10.9 Weathering classification for Mercia Mudstone

Class (after Spink and Norbury, 1993)		Example descriptions (Note – classes and example descriptions as BS 5930: 1999) *Colloquial descriptions from Chandler are not to be used*	Zone (Chandler, 1969)[1]
Reworked	E	Firm fissured reddish brown slightly gravelly CLAY. Gravel is fine to medium of subangular grey and yellow moderately weak sandstone. Rare rounded quartzite gravel. Fissures are subvertical extremely closely to closely spaced. *Solifluction deposit with foreign pebbles*	V
Destructured	Dc	Firm to stiff fissured reddish brown CLAY with rare subangular fine to medium gravel of grey moderately weak sandstone. Fissures are subvertical closely to medium spaced. *Clay matrix with no lithorelicts, but no pebbles*	IVb
	Db	Firm to stiff fissured reddish brown slightly coarse sandy slightly fine gravelly CLAY. Sand and gravel are of randomly oriented subangular to rounded lithorelicts of reddish brown very stiff clay to extremely weak mudstone. Fissures are subvertical generally medium spaced. *Clay matrix with <3 mm lithorelicts*	IVa
	Da	Firm to stiff reddish brown coarse sandy fine to medium gravelly CLAY. Sand and gravel are of randomly orientated subangular to subrounded lithorelicts of reddish brown very stiff clay to extremely weak mudstone. Occasional relict structure of randomly orientated discontinuities extremely closely spaced. *Clay matrix with frequent lithorelicts up to 25 mm*	III
Distinctly weathered	C	Firm to very stiff fissured reddish brown gravelly to very gravelly CLAY. Horizontal subangular to subrounded lithorelicts of extremely weak reddish brown mudstone. Fissures are generally subhorizontal and subvertical extremely to very closely spaced orange and black stained with up to 3 mm of clay or silt infill. *Lithorelicts of MUDSTONE in silt matrix*	III
Partially weathered	B	Extremely weak to very weak laminated to very thinly bedded reddish brown locally calcareous MUDSTONE. Subhorizontal and subvertical generally very closely spaced occasionally black stained discontinuities with a trace of silt or clay infill. *Angular blocks of unweathered MUDSTONE*	II
Unweathered	A	Weak to medium strong laminated to very thinly bedded reddish brown calcareous MUDSTONE. Subhorizontal and subvertical very closely to closely spaced discontinuities. *Unweathered MUDSTONE*	I

[1] Extended by Chandler and Davis (1973).

of the country where a concentration of development takes place. Chalk is special because it is generally a weak rock with a high porosity (up to 50% and typically over 25%). The high porosity causes it to remould readily under natural processes or the effects of sampling and this, together with weak calcite cement and narrow

Table 10.10 Weathering classification for Oxford Clay

Class (subdivisions after Spink and Norbury, 1993)		Example descriptions (Note – classes are as BS 5930: 1999) *Colloquial descriptions from Russell and Parker are not* *to be used.*	Zone (Russell and Parker, 1979)
Reworked	E	*Superficial materials of various origins, often flinty or* *chalky fragments in a clay matrix. Usually desiccated.*	IV
Destructured	D	*Not recognised.*	—
Distinctly weathered	C	Fissured brown CLAY with frequent fine gravel size gypsum crystals. Fissures with light grey gleyed infill *Totally oxidised CLAY, often with gleyed fissures. Fre-* *quent, often sugary, gypsum crystals*	III
Partially weathered	Bb	Bluish grey and brown CLAY with frequent fine gravel size gypsum crystals. *CLAY 30–60% oxidised around and beyond fissures.* *Large gypsum crystals*	IIb
	Ba	Very stiff fissured laminated bluish grey CLAY. Some sand size gypsum crystals. *CLAY is oxidised along surface of fissures only. Blue-grey* *colour and laminar fabric visible away from fissures.* *Some gypsum*	IIa
Unweathered	A	Very stiff dark fissured laminated bluish grey CLAY. Bedding horizontal. *'Firm' dark blue-grey silty CLAY. Some fissures. Horizon-* *tal laminar fabric*	I

pore throats, causes chalk to display fundamentally different behaviour from other weak rocks. In addition, the weak but stiff intact blocks are intersected by significant discontinuities which are often opened by dissolution.

Current practice for the description and classification of chalk is presented within the CIRIA report *Engineering in Chalk* (Lord *et al.*, 2002). This document incorporates guidance on description prepared by others such as Spink (2002) and Bowden *et al.* (2002). Before this time, industry practice in chalk description was based on the work at Mundford (Ward *et al.*, 1968) and extended by Wakeling (1970). This early site-specific study was the first attempt to provide a descriptive scheme tailored to chalk. The authors specifically asked for their scheme for this one site not to be used widely for other sites, not least because this site was in the Middle Chalk, whereas the majority of the outcrop is Upper Chalk; there are significant differences in the engineering characteristics of Upper and Middle Chalks. Nevertheless the Mundford system was widely used and abused, until the other papers cited and the CIRIA report finally ensured the earlier scheme was superseded. For reference, the Mundford gradings are included in the tables below but for completeness only. The Mundford gradings should not be used even if a detailed investigation is being carried out close to the Mundford site.

Because of the particular characteristics of the chalk, a programme of experimentation has found that description and classification can be most appropriately focussed on the density, which is loosely related to the strength of the intact material, and on the spacing and openness of the discontinuities within the structured chalk.

When chalk is subjected to disturbance, be this mechanical or thermal such as can be induced by freeze thaw action, it readily breaks down to form a structureless melange. The scheme for the description of chalk therefore has to include this condition. The structureless chalk will behave as a soil and two categories can be identified as being of importance – clast supported and matrix supported.

The two structureless chalk classes and the structured category where bedding and jointing are visible are summarised in Table 10.11, together with the particular word order that is adopted when describing chalk. This word order is the same as the general word order for all soils and rocks with chalk specific refinements. However, the principal soil type, and thus the word order to be used, depends on the category of chalk being described. Before starting it is therefore necessary to decide which sort of chalk is being described. This sounds more difficult than it actually is in practice.

The mechanical disturbance of chalk is most clearly seen in samples recovered using inappropriate techniques. The use of cable percussion methods with thick

Table 10.11 Categories of chalk for description purposes

Chalk type	CIRIA grade	Description word order	Mundford grade
Structureless fine soil	Dm	Structureless CHALK composed of: soil strength of matrix; colour of matrix; amount and size of clasts (sand and gravel); SILT; amount and size of clasts (cobbles and boulders). Nature of clasts (strength, density, colour, staining, angularity). Presence and nature of flints. Other features. (Stratigraphic member[1], grade Dm)	VI Putty chalk with small lumps (matrix supported)
Structureless coarse soil	Dc	Structureless CHALK composed of: amount and nature of matrix; angularity and size of clasts; GRAVEL; amount and size of clasts (cobbles and boulders). Nature of clasts (strength, density, colour, staining). Soil strength and colour of matrix. Presence and nature of flints. Other features. (Stratigraphic member[1], grade Dc)	V Putty chalk in comminuted matrix (clast supported)
Structured with bedding and/or jointing	C5–A1	Rock material strength; density; colour; staining of rock material; CHALK. Discontinuity type; discontinuity orientation (in situ observations only); discontinuity spacing; discontinuity aperture/infill thickness; nature of infill; discontinuity staining. Presence and nature of flints. Other features. (Stratigraphic member[1] and bed[1], density, CIRIA grade)	IV–I IV = Rubbly chalk III = Rubbly to blocky chalk II = Blocky chalk I = Brittle and massive chalk

[1] Stratigraphical unit name if known; this is not usually possible in structureless chalk unless the exposure is in situ.

walled driven samples is disastrous for sample quality. The only information that will come out of a borehole put down by these means is perhaps the level of the chalk surface, some SPT N values which are used in the design of pile base capacity, but are quite likely to be affected by disturbance ahead of the borehole and some flint fragments. No information will be forthcoming about the quality of the chalk in terms of fracture spacing and aperture except indirectly as the spacing will be at least as great as the largest lumps recovered.

Use of rotary coring produces much more representative exposure of the ground and is to be commended, but faces the problem that the driller is trying to recover one of the weakest rocks (the compressive strength of chalk is typically about 1–5 MPa) which contain one of the strongest rocks in the form of flint (the strength of flint is over 500 and up to 1000 MPa). Flints are often small but can extend to vertical heights of a few metres. Complete core recovery in these circumstances is a challenge, but is not impossible.

In situ examination of chalk is the best option, but man entry trial pits are limited to less than about 6 m depth by groundwater conditions. The amount of steel support put in place apparently to satisfy the health and safety rules severely reduces the amount of chalk exposed. Man entry inspection shafts are an excellent means of in situ logging, but again are severely restricted by practical and safety considerations.

The description of the chalk follows all the rules for rock description as set out throughout this book. The only addition in terms of description is that of density assessed in the field. The density should be assessed using the rock pick and nail tests provided in Table 10.12. The hand tests indicated should be carried out systematically as a cross check. The execution of the tests is illustrated in Figure 10.4. The spacing of the pick and nail tests can vary from site to site, but should generally be about every metre, no wider spaced than 2 m and only less than 0.5 m in exceptional circumstances. The published density tables include a correlation with compressive strength which is noted as being provided for information only as the correlation is not exact. This lack of correlation is particularly variable in the lower density or weaker chalks. The strength descriptors are omitted from the table here as their inclusion on the same table encourages loggers to assume a direct cross correlation which is a dangerous presumption. Loggers should assess the density using the tests in Table 10.12 and separately assess the strength using the terms in Table 5.7.

The description of the chalk should, wherever possible, record the discontinuity spacing and aperture. These parameters are crucial in the engineering behaviour of the chalk and are used as classifiers (Lord et al., 2002). Clearly, high quality samples or cores are needed for both of these aspects to be reliably recorded as noted above. The spacing and aperture classification terms are given in Table 10.13.

Particular care is required in assessing the aperture in rotary cores. Any movement of the core in the box could create an apparent aperture, or reduce an actual aperture, so particular attention should be given to measuring the total core recovery. In addition, the effect of flints on the core recovery and the nature of the fractures or any zones of core disturbance around flints and marl seams require the logger's care and attention. The reader is encouraged to examine exposures of chalk near the site, if available, or to view the examples of different chalk grades illustrated in the plates in Chapter 3 of Lord et al. (2002) as guidance on the appearance and character of in situ chalk.

Table 10.12 Field description of chalk density

Field identification	Rock pick penetration (mm)	Nail penetration (mm)	Density scale (intact dry density)
Chalk lumps (30–40 mm) crush between finger and thumb, remould to form putty.	>30 and splashes	>25	Low density <1.55 Mg m^{-3}
Chalk lumps (30–40 mm) can be broken in half using both hands but cannot be crushed between finger and thumb.	11–30	15–25	Medium density 1.55–1.70 Mg m^{-3}
Chalk lumps (30–40 mm) cannot be broken in half. Slabs <10 mm thick, or corners or edges of lumps (30–40 mm) broken with difficulty using both hands.	2–11	6–15	High density 1.70–1.95 Mg m^{-3}
Cannot be broken by hand. Lump (100 mm) can be broken by single hammer blow when held in hand.	<2	<6	Very high density >1.95 Mg m^{-3}

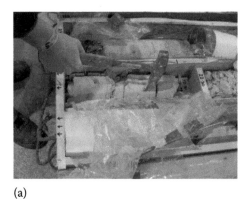

(a)　　　　　　　　　　　　　　　　　(b)

Figure 10.4 (a) Carrying out the rock pick test on chalk core and (b) measuring the pick penetration with a ruler. The penetration of a nail is measured in the same way.

The chalk classification scheme is then applied as density term/aperture grade/ spacing suffix, giving for example 'low density B4' or 'high density A2' as appropriate. Where the aperture or spacing varies laterally or vertically within a unit, it is suggested that the worst case classifier is adopted. The description of the exposure will, of course, have captured this variability, with description and classification being possible in more than one direction on the same exposure. There can be problems in assigning a grade when the evidence in the exposure is conflicting. For instance, if a closed discontinuity is stained, it is clear that water can pass through it. The question is then whether this is a grade A or B; many would go for the latter on grounds of conservatism. Without trying to resolve this argument here, full and accurate description should be carried out so that the reader can make up his or her own mind. The classification is less important and is a shorthand identifier to help others, not a critical element of the description.

Table 10.13 Chalk classification aperture and spacing terms

Typical discontinuity aperture	Grade
No infill in core	A
Closed in exposure	
Infill in core <3 mm	B
Open or infilled <3 mm in exposure	
Infill in core >3 mm	C
Open or infilled >3 mm in exposure	

Typical discontinuity spacing (mm)	Suffix
Spacing >600	1
200 < spacing <600	2
60 < spacing <200	3
20 < spacing <60	4
spacing <20	5

EXAMPLE DESCRIPTIONS (after Lord et al., 2002)

- Structureless CHALK composed of cream slightly gravelly sandy SILT with low cobble content. Gravel and cobbles are rounded extremely weak low density white with frequent black speckling. Occasional rinded flint cobbles. Locally light brown with a trace of brown clay. (Grade Dm)
- Structureless CHALK composed of slightly sandy silty subangular to rounded GRAVEL and COBBLES. Clasts are very weak low and medium density white with occasional black speckling. Matrix is light brown. (LEWES CHALK? Grade Dc)
- Very weak medium density white with frequent black speckling CHALK. Fractures very closely spaced (10/40/60), open and generally infilled (2/10/20) with light brown sandy silt size comminuted chalk, sometimes with a trace of brown clay, heavily brown stained. No discernible fracture sets. Horizontal band of medium spaced rinded flint cobbles at 2.70 m. (SEAFORD CHALK, CUCKMERE BEDS, medium density, Grade C4)
- Weak high density white, unstained CHALK. Fractures possibly medium spaced probably clean unstained. (RANSCOMBE CHALK, HOLYWELL BEDS? high density, Grade A?2?)

The example descriptions given above have been completed with the CIRIA grading included together with the geological formation. This is the same as the inclusion of weathering classifiers at this point of a description and is as presented in Lord *et al.* (2002).

This classification scheme encompasses all combinations of discontinuity spacing and aperture, unlike the previous Mundford scheme which only included those specific combinations that were found at the Mundford site. There was also a presumption in the Mundford scheme that the chalk quality improved with depth, which is known not to be the case in general. A comparison of the Mundford and CIRIA grading schemes is presented in Table 10.14 (Spink, 2002).

The CIRIA classification shows how a formation-specific scheme can be put together and which has been found to work very well in practice, giving engineers information that they can understand and which is applicable and useful. The Mun-

Table 10.14 Comparison of chalk classification schemes

| Discontinuity spacing (mm) | Discontinuity aperture | | | Grading scheme |
	Closed and clean	Open or infilled <3 mm	Open or infilled >3 mm	
Extremely close	M N/A	M N/A	M IV	Mundford
<20	C A5	C B5	C C5	CIRIA
Very close	M N/A	M III/IV	M IV	Mundford
20–60	C A4	C B4	C C4	CIRIA
Close	M II/III	M III	M III/IV	Mundford
60–200	C A3	C B3	C C3	CIRIA
Medium	M II/I	M II/III	M N/A	Mundford
200–600	C A2	C B2	C C2	CIRIA
Wide	M II/I	M N/A	M N/A	Mundford
>600	C A1	C B1	C C1	CIRIA

dford grading on the other hand was site specific, not formation specific, and was intended only ever to be applicable to that one site. This is the reason for the apparent gaps in the classes in Table 10.14 and illustrates the danger of extrapolating site-specific classifications across the outcrop without the due care that the original authors called for.

In the gap here between the description of chalk and karst, it is worth mentioning the description of dissolution features in a soluble rock such as chalk. These features are of very great significance to any engineering works where they are present and they should be described with due care. Detailed information is necessary on shape, or extent if encountered in a borehole, the nature and competence of any infill or the presence of any voids and any effects on the surrounding chalk in terms of weakening. Because of the colour contrast (see Figure 10.5) it is not usually difficult to spot their presence, but this may be obscured if inappropriate forms of investigation are in use.

Figure 10.5 Infilled dissolution features in chalk road cutting showing the complex geometry of such features (photograph courtesy of Digby Harman).

10.7.2 Karstic limestone

It can perhaps be debated whether karst dissolution is a weathering process or not, but given its importance to engineering projects in such terrain, it should be taken as such and thus included in the descriptive element of an investigation. It is apparent from the guidance given below from Waltham and Fookes (2003) that the description and classification scheme incorporates elements that will not be available from the description of a single core or exposure.

Because dissolution is a very different process to the weathering that affects most natural materials, a modified approach to the description of the weathering is merited. The features in a karst terrain that are important to the engineering, and are therefore used in the classification once the description has been made, include rockhead level and its variability and the size and spacing of any cavities or caves. A full description of the materials and other mass features such as discontinuities seen in any rock exposures or cores should be carried out in accordance with the guidelines and standard procedures set out in this book.

A description of the karst ground conditions should first state the rock type, whether this is limestone (or chalk) or gypsum. Four terms are then added to arrive at 'rock type/karst class/sinkhole density/cave size/rockhead relief'. The rock type is identified using normal geological terminology (see Chapter 4.11). Karst classification terms are given in Table 10.15.

The sinkhole density may be a number per unit area which can be derived from field mapping, maps or air photographs. The density could be related to the size so it should be noted if the density is low because the sinkholes are large. Ideally and if the necessary information is available, this descriptor should be presented as a rate of formation of new sinkholes (NSH), expressed in events per km^2 per year. In practice, such local records are rarely adequate for such an approach. It should be noted if the NSH rate is temporarily enhanced by engineering activities.

Typical cave size is a dimension in metres representing the largest cave width that is likely to be encountered. This is larger than the mean cave width, but may reasonably exclude dimensions of the extreme largest voids as these are probably rare, although these should be noted. This is an example of where use of a minimum/mode/maximum reporting of dimensions is useful.

Rockhead relief is a measure in metres of the mean relief in the karst rockhead, including depths encountered within buried sinkholes. Where possible, a note should distinguish between pinnacled rockhead and more flat jointed surfaces that represent buried pavements.

This four-element description and classification is considered the minimum that adequately represents the variability within an area of karst ground. The classification presented by Waltham and Fookes (2003) is summarised in Table 10.15. This approach provides an outline of the parameters selected for inclusion and, as in any classification, these identifiers should be taken as non-mutually exclusive. The reader is referred to the original paper for more and broader detail on the investigation of karstic terrain.

10.7.3 Tropical weathering

The approach to the description of tropical residual soils is essentially the same as for all other materials weathering from a rock through to a soil and is in accordance with the guidelines given in this book. There are modifications to the nor-

Table 10.15 Classification of karst

Karst class	Locations	Sinkholes	Rockhead	Fissuring	Caves
Juvenile KI	Only in deserts and periglacial zones, or on impure carbonates	Rare; NSH <0.001	Almost uniform; minor fissures	Minimal; low secondary permeability	Rare and small; some isolated relict features
Youthful KII	The minimum in temperate regions	Small suffusion or dropout sinkholes; open stream sinks; NSH 0.001–0.05	Many small fissures	Widespread in the few metres nearest surface	Many small caves; most <3 m across
Mature KIII	Common in temperate regions; the minimum in the wet tropics	Many suffusion and dropout sinkholes; large dissolution sinkholes; small collapse and buried sinkholes; NSH 0.05–1.0	Extensive fissuring; relief of <5 m; loose blocks in cover soil	Extensive secondary opening of most fissures	Many <5 m across at multiple levels
Complex KIV	Localized in temperate regions; normal in tropical regions	Many large dissolution sinkholes; numerous subsidence sinkholes; scattered collapse and buried sinkholes; NSH 0.5–2.0	Pinnacled; relief of 5–20 m; loose pillars	Extensive large dissolutional openings, on and away from major fissures	Many >5 m across at multiple levels
Extreme KV	Only in wet tropics	Very large sinkholes of all types; remanent arches; soil compaction in buried sinkholes; NSH >>1	Tall pinnacles; relief of >20 m; loose pillars undercut between deep soil fissures	Abundant and very complex dissolution cavities	Numerous complex 3D cave systems, with galleries and chambers >15 m across

mal approach that are necessary because of the particular effects of the weathering being in situ and largely chemical. Full details are given in Anon (1990) and are summarised below.

The descriptive scheme starts with a number of details such as site location, landform details, bedrock geology, current climate, topography, hydrology and vegetation. These are outwith the actual description of an exposure or sample and so are not included here. In terms of the actual description of the ground, the suggested approach for the description of soil material is summarised in Table 10.16.

The soil mass should be identified and described to provide information on the distribution of material and weathering types in the profile. Four aspects of the mass character are identified as being necessary to provide the required information.

- Composition is to include the distribution and character of the various materials which may require sections or scan lines

Table 10.16 Descriptive scheme for residual soil materials, after Anon (1990)

Parameter	Classification	Procedure
Moisture	State	Dry = light colour, loose brittle
		Moist = range of colours, neither wet nor dry
		Wet = visible water films
Colour (see Chapter 7)		Use colour charts and describe as for other soils or rocks
Strength (see Chapter 5)	Fine soil	Describe consistency
	Coarse soil	Describe relative density
	Rock	Describe rock strength
Fabric (see Chapter 6)	Origin	Orthic = formed in situ by soil forming processes, e.g. coatings, nodules, peds
		Inherited = relicts of parent material, e.g. lithorelicts
	Voids	Low, medium and high terms based on laboratory measurements of porosity
	Orientation (See Table 2.4)	Strong = most particles are sub parallel
		Moderate = many particles are sub parallel
		Weak = some particles are sub parallel
		None = no particles are sub parallel
	Distribution	Porphyritic = matrix is dense
		Agglomeritic = matrix loose or incomplete
		Intertextic = grains embedded in porous matrix
		Granular = no groundmass
	Fissuring	Orientation, spacing and character as per standard guidance
Texture (see Chapter 6)		Grain size as per standard guidance
Density		Low, moderate and high terms based on laboratory measurements of density or relative density
Apparent behaviour	Remoulded strength	As density above
	Durability	Slakes, breaks, chips terms based on field tests
	Plasticity	Plasticity described as per standard guidance
Mineralogy		Hand lens examination, carbonate content test

- Structure should include any naturally occurring boundaries between materials and any contemporary or relict discontinuities including bedding, jointing and faulting (see Table 10.17) which gives the terms for description of boundaries and contacts.
- Discontinuities should be described in accordance with the guidance given in Chapter 11 of this book.
- Behaviour should be recorded with respect to the effect of conditions imposed during the proposed construction. This will require field tests such as dynamic penetration tests, vane shear tests, cone penetration tests, pressure meter or dilatometer tests, CBR tests, plate bearing tests and in situ permeability tests.

The field identification procedures for the material and mass outlined above and provided in detail in Anon (1990) will form the first judgement about the ground type and performance. This initial assessment will necessarily be modified and refined in the light of subsequent laboratory and field testing.

Classification of tropical residual soils requires a specific approach relevant to the weathering processes present in the profile. Most engineering geological classifications are developed to describe the behaviour of soils formed under temperate or colder climatic conditions. These soils are often little altered sedimentary or transported deposits and are based on some combination of fabric features, plasticity and particle size grading. The engineering behaviour of tropical residual soils is rather more difficult to predict because the weathering products contain different minerals from the host rock and these may have unusual properties and because in situ weathering suggests the retention of some real or relict in situ structure which may persist and have a significant effect even when weathering is very advanced. For details of the classification schemes, the reader is referred to Anon (1990).

Table 10.17 Terms to describe contacts and boundaries

Boundary distinctness		
Class	**Boundary thickness**	**Term**
1	<5	Sharp
2	5–25	Abrupt
3	25–60	Clear
4	60–130	Gradual
5	>130	Diffuse
Boundary form		
Class	**Description**	**Term**
1	Boundary form is a plane with few or no irregularities and is usually at the same depth across the exposure.	Smooth
2	Boundary has broad, shallow, regular pockets.	Wavy
3	Boundary has pockets which are deeper than the width.	Irregular
4	At least one of the horizons is discontinuous and the boundary is interrupted.	Broken

11 Discontinuity Logging

11.1 Types of discontinuity

The logging of discontinuities (or the synonymous term fractures) is of crucial importance in soils and rocks. Blocks of intact material, which are normally those tested in hand tests, are broken by discontinuities into a fractured mass which displays different properties from the material. The discontinuities greatly affect the strength, compressibility and permeability of the ground. The presence of discontinuities in soil is very important, but the discontinuities are of overriding importance in rocks where the reduction from material to mass strength or compressibility is proportionally greater.

Description of a soil or rock mass can not therefore be complete without a thorough description of the fractures present, their spacing and orientation and their character. The approach and terms used to describe these features are outlined in the following sections. The approach to carrying out field logging of discontinuities in terms of face mapping and the use of scan lines is considered in Chapter 17.

The various types of fractures that will be encountered and that therefore require description have been defined in the Preface and include:

- bedding planes
- bedding fractures
- cleavage and cleavage fractures
- joints (in rock) and fissures (in soil)
- shear surfaces
- faults, fault zones and fault gouge
- incipient fractures and
- induced fractures.

Most of the above fracture types are those that can be expected to occur naturally in the ground and whose correct description characterises the in situ condition. This characterisation is a requirement of BS 5930: 1999 Clause 47.2.6.1. This clause requires the field description to report on the ground conditions at the position of the exploratory hole before being subjected to loss and disturbance by the drilling or excavation process. It is therefore a fundamental requirement of fracture description to identify and exclude from the assessment those fractures that are caused or created by the formation of the exposure, that is the incipient and induced fractures. These disturbing processes can include core drilling (and subsequent handling or deterioration), excavation or blasting.

Induced fractures could be considered for inclusion in the description and could be included in a fracture assessment if a non-standard approach is followed. There are considerations, such as ease of excavation and material re-use which may benefit from this information. However, one of the reasons for the BS 5930 requirement is to produce descriptions of the ground that are not a function of the quality and care exercised in the drilling, sampling or excavation process.

The range of terms in use for different types of fracture is not standardised in the context of a normative standard. The definitions used here (see Preface) were introduced in BS 5930: 1999 in an attempt to provide more consist terminology and has been found to work well. In practice, the definition of types of fracture can follow local practice and the only essential requirement is that the terms used and their definitions are understood by the reader and that they follow conventional geological practice in nomenclature.

11.2 Discontinuity description

Discontinuities within the soil and rock mass are of such primary importance to the overall engineering properties that the maximum possible amount of information must be reported on all occasions when such features can be identified. It should be noted that appropriate samples should be taken to allow this description to be made (in accordance with EN ISO 22475-1). This means that if it is suspected that the discontinuities will be critical to the investigation of the engineering problem, high quality (Category A, or possibly Category B) samples should be taken from boreholes. Even better, in situ exposures should be created to allow the features to be recorded in a less disturbed condition and identification of their actual characteristics and orientation in three dimensions made. Appropriate trial excavations could include pits, trenches, shafts or adits, depending on the ground conditions and the nature of the engineering problem for which the investigation is being carried out (see Chapter 17).

The information that should be recorded in logging discontinuities is listed below. Terms and guidance on their use are given in the order in which this should be presented. Field proformae for recording this information should be laid out in the same sequence for convenience and accuracy and some example proformae are presented in the Appendix:

- orientation
- spacing
- persistence
- termination
- roughness
- wall strength
- aperture
- infilling
- seepage
- number of sets.

Guidance on the description of all these aspects is given in EN ISO 14689-1 and BS 5930: 1999 and includes the provision of quantitative classifier terms for spacing, persistence, termination, roughness, aperture and seepage. The descriptive requirements are set out below in their word order sequence.

TIP ON DISCONTINUITY LOGGING IN BOREHOLES

The amount of information available for description of discontinuities in borehole cores is limited, as the core represents only a one dimensional scan line. Those elements of discontinuity description that can be carried out on a borehole core are normally limited as follows:

- Orientation – only the dip amount can be given unless some form of core orientation has been carried out.
- Spacing of the flatter discontinuities can be provided, but the true spacing of steep or vertical fractures cannot be given.
- Roughness terms at small and possibly intermediate scale can be given.
- Wall strength can be described although it may be difficult to distinguish natural from induced characteristics.
- Aperture can be reported but care is required that the reporting is not merely representing movement of core pieces in the box.
- Infilling can be described in terms of extent and nature.
- Number of sets can be indicated at least in outline.

11.3 Orientation

Orientation is measured in terms of dip direction and dip. Historically the concept of strike was used instead of dip direction, but this is no longer favoured, at least in engineering geological practice. The reasons for this include the absence of any ambiguity when measuring dip direction rather than strike (which always has two possible dip directions). It may be that the overriding reason for this change is rather more prosaic in being due to the use of spreadsheets to record and manipulate data, in which a further non-numeric field is required to record the direction of the dip.

The dip direction and dip, defined in Figure 11.1, are measured relative to the horizontal and reported as three figure azimuth or dip direction and two figure dip amount, thus 050°/25°, for example. If this convention is followed, the sequence is not critical as it will be self evident which reading is which; however care should be taken if the data is printed from a spreadsheet in which case any initial zeros will be

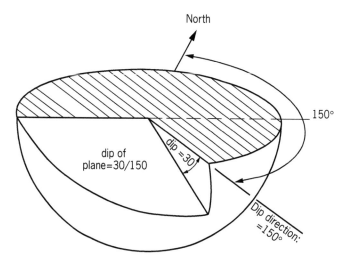

Figure 11.1 Definition of dip and dip direction in accordance with standard nomenclature.

lost. The assessment of orientation in cores is obviously limited as it is not possible to measure the dip direction unless other measures have been taken to orientate the core. This can be done by orientation of the core from the hole using a core orientation device (of which there are several choices on the market), or by using down hole logging in some form.

The dip can be measured, however, and in order to accord with standard practice, this is measured as the dip relative to the normal to the core axis, that is to the horizontal. The dip should be measured with a clinometer or protractor and can be readily measured to the nearest degree or so. Given the inherent variability in the dip on individual natural surfaces, it is often reasonable to consider reporting dips to the nearest 5°, but if the readings are taken to the nearest degree, they should be reported as such. The reporting of dips to anything coarser than 5° is not considered acceptable.

TIP ON DIP MEASUREMENT IN INCLINED BOREHOLES

- In cored boreholes that are other than vertical the convention that the dip is measured relative to the normal to the core axis should be followed.
- It must be recognised that the reported figure is not the dip in the ground. Inclined boreholes will need to include some orientation of the borehole and the core to allow the actual dip to be determined.

TIP ON REPORTING EXTREMES OF DIP

- There is a commonly held view that the dip of steep fractures can be summarised by use of the term subvertical (80°–90°) and the dip of shallow fractures by the term subhorizontal (0°–10°). The use of these terms is not encouraged.
- The reporting of actual dip measurements in degrees is encouraged even within these ranges.
- The reason for this preference is that the use of summary terms may encourage the dips not to be measured. In many circumstances, the presence of dips up to 5° from the horizontal (or vertical) can give important indications of structural features such as faults in some proximity to the investigation hole.

One aspect of the use of compass clinometers in the field that is not covered by text books is the use of the zero adjustment screw to take account of magnetic declination at the site location. This situation is made worse because many undergraduate field leaders encourage this practice, not necessarily with consistent outcomes. In these days of simple manipulation of collected data in a spreadsheet, the possibility of error means that adjustments of this kind should not be made in the field.

It is imperative to record on field sheets if any north reference other than magnetic is used, for instance if readings are reported to a site grid or some other reference. Leaving this information to be reported on some key sheet or, worse, in the report text, is not satisfactory as reports tend to get dismembered with time such that field logs and covering key sheets or text do not remain together.

(a)

(c)

(b)

Figure 11.2 (a) Compass clinometer (photograph courtesy Silva AG) and a clinometer in use (b) to measure the orientation of a joint in core and (c) a shear surface in situ.

The measurement of the orientation of planar features is straightforward where the surface is indeed planar. However, most natural discontinuity surfaces in soil and rock are far from planar, an element of variability that is dealt with by the description of surface form below. There are a number of different approaches to taking representative readings in such circumstances. Use of a variety of plates of different sizes enables sampling of the average attitude to be made at a range of different scales (ISRM, 1981). Alternatively, a large number of readings can be taken over the exposed surface area in order to represent the variability. However, opportunities where there is sufficient exposure of the joint surface are rare. The best approach that would normally be used in engineering geological practice is to take random readings which can be assumed to be where the discontinuity intersects the scan line. If it is considered that the discontinuities vary significantly in orientation and that this would not provide sufficient information on the variability, additional scan lines should be carried out at close intervals.

TIP ON RECORDING ORIENTATION IN THE FIELD
- An approach which is favoured by many experienced field geologists would be to stand back and take an overall reading of the discontinuity orientation from the whole of the information available. This approach allows sensible assessment of the overall discontinuity attitude rather than numerical rigour and is often appropriate, given the extent of exposure that is available in many investigations.
- This overall reading would then be complemented by the description of the surface form at three scales, given in Chapter 10.6.
- However, this approach does require careful thought in the field and in any subsequent use of the data about the size and orientation of blocks in the ground as a whole. This is thus not an approach that can readily be given to inexperienced field loggers, for whom the more extensive and rigorous approaches outlined above are likely to be more appropriate.

11.4 Spacing

The spacing of discontinuities controls the size of blocks within the rock mass and thus the overall behaviour in slope failures and excavation, for example. The spacing that is to be described is always the separation, or normal spacing measured at right angles to the planar surface. This should be recorded as actual measurements in millimetres or can be summarised using the classification terms in Table 11.1. In exposures, this is straightforward and presents no particular problems. In borehole cores however, only the discontinuities normal to the core axis are fully sampled by

Table 11.1 Terms for describing spacing of discontinuities

Spacing (mm)	Term
over 2000	Very wide
600–2000	Wide
200–600	Medium
60–200	Close
20–60	Very close
under 20	Extremely close

TIP ON CLEAVAGE
- The terms in Table 11.1 should not generally be used for cleavage which is a penetrative foliation which does not have spacing; the cleavage fracture spacing will change with time and core condition, and so the repeated spacing needs a date or note on the state of the core.

the core sample. As in field exposures these can be readily described. The description of the spacing of inclined fractures needs to take account of the in situ geometry even though there may not be many fractures visible in the core.

The spacing descriptors should be given for the most common classes. It is not necessary for the terms to cover the whole range just in order to include the rare extreme readings. The presentation of descriptions with too wide a range of spacing terms is not helpful. It is preferable to give the actual range of spacing or thickness as a measurement in addition to the use of the descriptors. The measurements given can include the extremes. This approach allows the descriptor to be helpful, without causing undue concerns arising from not including the extremes. This is illustrated in the examples below.

EXAMPLE DESCRIPTIONS

- Very closely fissured
  Minimum/typical/maximum spacings are 15/40/75.
- Generally closely spaced desiccation cracks
  Minimum/typical/maximum spacings are 25/100/220.

TIP ON RECORDING STEEP FRACTURES

- The spacing of those fractures which are parallel or near parallel to the core axis cannot be determined in cores, as they will be significantly under sampled. It is possible for core to be recovered with more than one vertical fracture present, in which case the spacing can be measured (see Figure 11.3). Nevertheless, caution needs to be exercised in the description of the spacing of fractures parallel to the core if the reader is not to be misled.
- The only measurement that can be made is that of spacing along the core axis. Although a useful record of what is in the core, such a record is of little use in aiding the reader of the log to make their assessment of the rock mass in situ.
- Given the bias to the sampling of vertical fractures (as most boreholes are drilled in a vertical direction) each fracture should be reported individually on the field log, in order that the reader can make their own assessment.
- These individual fracture records should include top and bottom depths, thereby providing an indication of frequency, axial spacing and persistence.

Figure 11.3 Chalk core containing at least three vertical discontinuities, allowing the spacing to be described, even though this may still not be representative as the sampling of axis parallel fractures is not usually complete.

EXAMPLE OF EFFECT OF STEEP FRACTURES

A site investigation was carried out for a new sewer tunnel in Coal Measures in Lancashire. It was realised that every piece of core included at least one vertical fracture. Given that vertical cores do not fully sample vertical fractures, this was clearly unusual and indicated a high incidence of vertical fractures. With this realisation in mind, the interpretative report highlighted that roof stability was likely to be problematic in the tunnel. This turned out to be the situation in fact and thus potential problems were foreseen.

Measurement of the spacing in logging in situ exposures can be done in two ways. Either the approach described above for cores can be followed, with individual measurements being made on whatever discontinuities are available. Alternatively and preferably where exposure is available, a more systematic approach can be adopted. This is the approach preferred in ISRM suggested methods (ISRM, 1978). A base line or scan line should be established and measurements made of the individual fractures crossing the line. The measurements can then be reported individually or averaged to give a spacing for each set. A minimum length for this measurement should be about 3 m, but longer is preferable, ranging up to 10 m. The measurement of spacing along three orthogonally arranged scan lines is even better, permitting the results to be analysed in three dimensional space, see Chapter 17.

11.5 Persistence and termination

Persistence is the areal extent of a discontinuity. This normally requires the extent of the discontinuity surface to be measured both down dip and across dip. In practice, this is rarely feasible and the measurement is normally the linear extent of the discontinuity trace that is visible in an exposure. The persistence is a very significant factor in the assessment of a rock mass and its behaviour, but it is also difficult, not least because exposures are rarely large enough for a comprehensive set of observations to be made.

Classification terms for persistence are provided by ISRM (1978) and are as given in Table 11.2. These terms do not replace recording of actual persistence in the field, but can be useful when describing the rock mass in a summary description of an exposure or in the narrative text of a report.

Persistence terms are not often applied in cores because the small extent of the exposure does not allow their determination except perhaps for vertical fractures.

It is also important to record the type of termination of the discontinuity. This is commonly the situation where one set of discontinuities terminate against another set, either completely or by displacements which can be large or small. Terminations

Table 11.2 Terms for discontinuity persistence

Persistence (m)	Term
<1	Very low persistence
1–3	Low persistence
3–10	Medium persistence
10–20	High persistence
>20	Very high persistence

can either be beyond the exposure (x), against another discontinuity (d) or within the rock material (r). The termination of discontinuities is not usually identifiable in cores.

11.6 Surface form

The roughness or smoothness of a discontinuity surface is a crucial component of the shear strength of the discontinuity. Asperities on the surface of the discontinuity mean that for shearing to take place, the two sides have to ride over or shear through the asperity and thus the strength is higher than if the asperity was not present. The influence of the roughness is further modified by the wall strength and any infilling, the description of which is covered below.

The description of the surface character of discontinuities is given at three scales. Large is at a scale of metres, medium is at a scale of centimetres and small is at a scale of millimetres as indicated in Table 11.3. Typical surface profiles for use at each of the three scales are given in Figure 11.4. The set of profiles should

Table 11.3 Terms for description of surface form of fissures, joints or other discontinuities

	Small scale (mm)	Medium scale (cm)	Large scale (m)
Length of profile on Figure 11.3	10 mm	100 mm	1000 mm
Profile number			
1–3	Striated (in direction of striae)	Planar	Straight
4–7	Smooth	Undulating	Curved
8–10	Rough	Stepped	Wavy

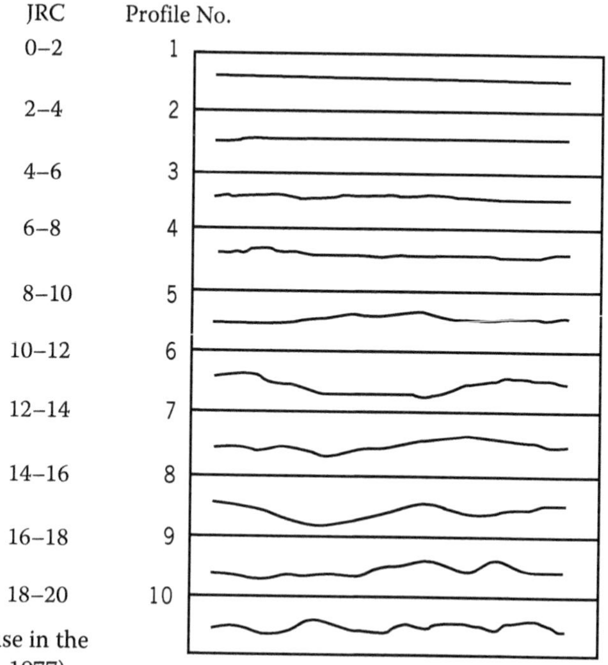

JRC	Profile No.
0–2	1
2–4	2
4–6	3
6–8	4
8–10	5
10–12	6
12–14	7
14–16	8
16–18	9
18–20	10

Figure 11.4 Joint profiles for use in the field (after Barton and Choubey, 1977).

be printed out at appropriate sizes to give the three lengths indicated in Table 11.3, ensuring that the enlarged profiles have retained their aspect ratio.

It is apparent that description at the small scale only can be available in cores, although some description at the medium scale may be possible depending on the orientation of the fracture relative to the core. These scales are those that have been used for many years in international practice (ISRM, 1978). However, the terms defined in BS 5930: 1999 were at an order of magnitude larger than this (decametres, metres and centimetres). It is not clear whether practitioners who were nominally describing discontinuities in accordance with this Standard actually realised this difference. It appears that most UK practitioners have actually been following the ISRM practice in ignorance.

The surface form of discontinuities should be recorded as a matter of course both in soils and rocks, be these fissures, shears or joints or any other form of discontinuity. The level of detail that is included in everyday descriptions is normally modest. However, those cases where the effect of the discontinuities really matters to the mass strength of the ground will merit greater detail being recorded.

Additional detail can also usefully be provided to indicate whether the surfaces are matt or polished. Polished surfaces can be described as slightly polished, polished or highly polished, such terms being relative for general use. If a surface has signs of shear movement, in the form of linear stripes or asperities, the term striated is preferred to slickensides for the reasons given below. Surfaces that have undergone substantial movement can display significant striations at a range of scales. Full description of such surfaces can be best achieved by describing the surface form both parallel and perpendicular to the striations. The strike direction of the striations is also useful information that should be recorded together with any evidence about the direction of movement. This evidence can be visible in the exposure or inferred from the setting or other information.

TIP ON USE OF DISCONTINUITY TERMS

There are a number of terms in use for the description of discontinuities which are not preferred, as these are terms whose meaning is not clear and so potentially ambiguous. These terms have come either from the miners who are the prime amateur geologists or from other areas of the geological dictionary.

- A slickensided surface is a surface on which movement has taken place giving rise to striations and polish. Dictionaries and the geological literature contain a variety of definitions and examples. Some of these indicate the presence of high polish, but others do not, some have lineations, others do not. It is therefore better to avoid any confusion and to describe the surface form, striations on the surface and also separately to describe the polish.
- Listric surfaces are literally spoon-shaped surfaces. The term is used at a wide variety of scales from features a few centimetres across to major faults. The terms given above provide for the description of such morphologies using defined terms. The term listric should not therefore be used in a description.
- A greasyback is a miner's term for the description of various styles of separation surfaces in mudrocks and was widely used in London Clay. Terms for the description of such surfaces are provided above and the use of undefined terms which can be confusing is not encouraged.

EXAMPLE DESCRIPTIONS
- Stiff fissured grey CLAY. Fissures are randomly orientated very closely spaced (20/50/65) wavy planar smooth matt.
- Firm sheared yellow brown CLAY. Shears dip 10/150–170 are medium spaced (150/250/300) persist up to 500 mm and are straight planar smooth polished and locally striated.

The terminology for description of discontinuity surfaces includes a number of miners' terms which have grown up over many years. These include 'greasybacks' or 'listric' surfaces. Whilst there is nothing specifically wrong with the use of these terms, they are poorly defined and their use can be ambiguous or unclear. Restriction of the description of the discontinuity surfaces using the terms and quantitative terminology given above is preferred.

11.7 Wall strength

The strength of the soil or rock forming the discontinuity walls is an important component of the shear strength of the discontinuity. This is especially the case in unfilled discontinuities where the walls are in direct material to material contact. In these cases, the condition of the wall may have more effect on the shear strength than the material behind the discontinuity wall. For example, if the wall is weathered or degraded the wall strength can be a small fraction of the original strength in the interior of the blocks and the overall shear strength of the material will be reduced accordingly.

The strength of the material forming the wall of the discontinuity should be determined if possible. This sounds easy, but is in practice difficult because the wall material tends only to comprise a thin skin. Although the Standards indicate that the strength can be measured, such as by vane or Schmidt hammer tests, this assumes there is sufficient thickness of material to test. This is commonly not the case and the test procedure may actually be measuring the strength of the stronger material behind the discontinuity.

The description of the strength of discontinuity walls in cores and borehole samples needs to be carried out with caution, given the softening effects of drill flush or borehole fluids on these materials. The walls of discontinuities in exposures can also change after creation of the exposure, either because of ground water flowing out through the discontinuity, action of rain on the face or other weather effects on the face, such as drying out.

TIP ON MEASURING WALL STRENGTH
- The logger is well advised to ensure that a field assessment of the strength is made using the hand tests and descriptive terms given in Table 5.8. The description of the loss of strength should include the degree of loss, as well as the extent of the effect in terms of its penetration into the material block. This information is part of the description of weathering effects, see Chapter 10.

11.8 Aperture and infilling

If the discontinuity is open, the aperture may or may not be infilled. If there is no infilling, groundwater can flow freely. Even discontinuities only open a few millimetres can transmit significant quantities of water with detrimental effects on tunnelling machines or slope stability. Alternatively, the discontinuity infill can be a weaker material such as weathered product or fault gouge. This infilling material will reduce the permeability and will also have a significant effect on the shear strength of the rock mass, again with a detrimental effect on strength and thus tunnel or face stability. There are also partially infilled discontinuities which are intermediate between these two extremes. It is therefore apparent that the openness and infilling of the discontinuity are significant aspects of the description.

The aperture is the perpendicular distance separating the walls of a discontinuity and this space can be filled with air, water, soil or rock. In most rock masses, individual discontinuities are small with apertures of less than a millimetre to a few millimetres. Much larger apertures are, of course, possible where flowing water has eroded joint wall material, or dissolved the rock material.

The terms provided for description of discontinuity aperture are given in Table 11.4. As with all aspects of discontinuity logging, these terms are to assist in communicating the results to others and do not replace accurate measurement and recording in the field.

The aperture of most discontinuities is not constant, but will vary, not only along the trace length but also with time depending on stress conditions. The current condition of the discontinuity at exposure or in the core box is not subject to the same regime as the discontinuity has been in the past or future when in the ground. It is possible for joints that are tightly closed to be stained, as the in situ stress condition could have caused opening previously. Similarly, the aperture of joints can change when affected by the proposed works such as excavation or tunnelling. Variations in the aperture that can be discerned at the exposure should be recorded and this information is of critical importance to allow the reader to make an assessment of possible changes to discontinuity parameters in situ. The relevant field information will include features such as staining or infill.

Particular care is required in assessing the aperture in rotary cores. Any movement of the core in the box could create an apparent aperture, or reduce an actual aperture, so particular attention should be given to measuring the total core recovery.

One of the great difficulties in assessing aperture is that the very process of creating the exposure will probably have caused some disturbance, generally resulting in some opening of the discontinuity. Accurate logging requires the effects of such

Table 11.4 Terms for describing discontinuity aperture

Aperture spacing (mm)	Term
< 0.1	Very tight
0.1–0.25	Tight
0.25–0.5	Partly open
0.5–2.5	Open
2.5–10	Moderately wide
10–100	Wide
100–1000	Very wide
> 1000	Extremely wide

disturbances to be discounted and the original in situ condition to be assessed and described.

This difficulty is particularly marked in the assessment of the discontinuity aperture in recovered cores. Even small amounts of disturbance to the core will result in the recovered condition not being representative of the in situ condition. The only two cases where the aperture can be reliably reported in core logging are where the discontinuity is tightly closed or where it is completely infilled and the infill is also recovered to the core box. On all other discontinuities, even slight movements of the core can cause discontinuities to open or close by significant amounts. It should be possible to use the total core recovery (see Chapter 12) as an aid in assessing the discontinuity aperture. However, the recording of core recoveries and depths is not always sufficiently accurate in combination. The application of all core loss to fracture openness is likely to result in a cautious estimate of the rock mass quality. Whilst it could be argued that this caution could be appropriate, there is a real chance that the assessment will be too pessimistic. Loggers should consider these matters in context at the core box.

The infilling of discontinuities can comprise a wide range of materials, including material weathered in situ, minerals grown in place and material imported by groundwater or hydrothermal fluids. The accurate description of all infilling materials is important and should follow the rules for description of soil or rock material as relevant and presented throughout this book. In particular, details of the infill thickness, strength, whether it is soil or rock, colour, mineralogy, the particle size grading, fracture state and any signs of shear displacement should be included.

11.9 Seepage

The identification and description of seepage is important in order to enable designers to identify the secondary permeability of a rock. Most water flowing through rock moves through the fractures but can flow through the material in, say, a coarse soil. The description of seepage, which can clearly only be made on field exposures, should identify the amount of water flowing and whether this is taking place through the material or through discontinuities. If the flow is through discontinuities, the relevant discontinuity or set should be identified (see below). The description of seepage as outlined below is a much safer approach than the approach outlined in early Working Party reports (Anon, 1977) where the field description included an assessment of permeability.

Seepage from discontinuities in the face can be estimated simply using a bucket or other sized container to calibrate the rate of flow. A simple classification is available in EN ISO 14689-1, given in Table 11.5. Again this is for communication and ease of reporting and does not replace specific field measurements. It is apparent

Table 11.5 Terms for description of seepage

Rate of flow (litres s^{-1})	Term
0.05–0.5	Small
0.5–5	Medium
>5	Large

that the seepage from discontinuities cannot be described in cores recovered to the surface.

Field records of seepage should include information on other factors which could affect the seepage, such as other nearby construction activities including pumping, time elapsed since face excavation and recent weather conditions.

11.10 Discontinuity sets

The overall mechanical behaviour of the rock mass will be determined by the number, arrangement and intersection of the sets of discontinuities that are present. Depending on the spatial arrangement, the rock mass could display marked anisotropy with respect to particular orientations of engineering works. The stability of rock slopes on either side of a road cutting will vary markedly as a result of the slope orientation with respect to the discontinuities as will the stability and stand up time of an underground excavation. In addition, each discontinuity set could have different properties, described in the preceding sections, not least in terms of orientation and spacing.

The arrangement of multiple sets of discontinuities is not usually determinable in cores other than in outline. However, it is perfectly reasonable to expect that a logger will be able to distinguish the shallow dipping, inclined and steep sets of fractures and to log these individually as illustrated below. Of particular importance is recording any changes in the pattern of fracture sets and these should be clearly distinguished. In practice, the zones of particular styles or combinations of fracturing will generally result in a number of separate strata on a field log.

Further detail on the sets of joints in a rock mass should be given when logging in situ exposures. The descriptive scheme given in Table 11.6 for the number of

---

EXAMPLE DESCRIPTION

Description of sets of fractures in a core with actual measurements given in mm
- Joints subhorizontal parallel to bedding, very closely spaced (15/20/45) planar smooth stained. Joints dipping 45° are closely spaced (50/80/120), undulating rough lightly stained. Vertical joints are planar smooth to rough clean.

Table 11.6 Scheme for the description of the number of joint sets

Classifier	Number of joint sets
I	Massive, occasional random joints
II	One joint set
III	One joint set plus random
IV	Two joint sets
V	Two joint sets plus random
VI	Three joint sets
VII	Three joint sets plus random
VIII	Four or more joint sets
IX	Crushed rock, earth like

Table 11.7 Description of fracture bounded rock blocks

Term	Shape of blocks
Polyhedral	Irregular discontinuities of short persistence without arrangement into distinct sets
Tabular	One dominant set of parallel discontinuities; thickness of blocks much less than length or width
Prismatic	Two dominant sets of generally orthogonal discontinuities; thickness of blocks much less than length or width
Equidimensional	Three dominant sets of approximately orthogonal discontinuities; Blocks are equidimensional
Rhomboidal	Three or more dominant sets mutually oblique; oblique shaped equidimensional blocks
Columnar	Several discontinuity sets (more than three) crossed by irregular joints; length much greater than thickness or width

joint sets was proposed by ISRM (1981). In addition to this scheme, major through going fractures had to be recorded individually.

This sort of approach has been extended in other standards (IAEG, 1981; EN ISO 14689-1) with the number of joint sets and their interrelationship given in terms of orientation, see Table 11.7. This approach provides an indication of the shape of the fracture bounded blocks.

The text of all standards provides diagrams to illustrate what is meant by each of these terms. These diagrams are not repeated here as they are unlikely to be useful in field practice. The use of terms such as those outlined in Table 11.7 should be restricted to narrative texts and are not part of a descriptive approach that would be practical or advantageous for use in the field.

Example descriptions of discontinuities in a soil and a rock are given below, using the terms summarised in the preceding paragraphs.

TIP ON RECORDING SHAPE AND SIZE OF JOINT BOUNDED BLOCKS

- On those occasions when there is sufficient exposure available for any of the given descriptors of the shape and size of blocks in three dimensions to be used, the most appropriate approach would instead be to carry out a systematic logging exercise, see Chapter 17.
- This would entail the establishment of preferably three or more orthogonal scan lines and use of these to record the relevant details for each intersecting discontinuity using the terms and in the order described in the immediately preceding paragraphs.
- This information can then be presented in summary tables and manipulated in spread sheets and the orientation can be presented on stereonets or rose diagrams as preferred. The results of such a field approach would then also provide useful data to input to a rock mass classification scheme such as the rock mass rating (RMR) or Q system (Bieniawski, 1989).

EXAMPLE DESCRIPTIONS

- Stiff fissured sheared greyish brown CLAY. Fissures are horizontal very closely spaced (15/45/60) persistence >100 mm planar smooth with low polish. Fissures are vertical persistence up to 500 mm undulating rough stained yellowish brown. Shears are 40–50° medium spaced (250/300/500) planar smooth highly polished and striated.
- Strong light and dark orange brown and moderately strong light grey medium grained SANDSTONE (CARBONIFEROUS SANDSTONE). Joints dipping 10 to 15°, closely spaced (80–120 mm), planar smooth stained. Joints dipping 45°, very closely spaced (40–60 mm), undulating rough stained. Vertical joints smooth to rough clean.
- Firm sheared and fissured grey mottled orangish brown slightly gravelly CLAY. Gravel is fine and occasionally medium of grey mudstone peds (destructured MUDSTONE). Shears are medium spaced, persist for 0.5–>3 m, light grey gleyed with slight polish dipping 350–360° at up to 8°. Fissures are subvertical with random dip direction, brown and light grey gleyed, straight undulating and rough.
- Very stiff to extremely weak dark grey occasionally mottled orange thickly laminated CLAY/MUDSTONE (distinctly weathered MUDSTONE). Joints dip 5–10° at 225–245°, persist >2 m, grey with occasional light grey staining, wavy, undulating and smooth. Joints are subvertical dipping to north-south and east-west grey and brown mottled, straight, planar, rough.

12 Fracture State Recording

Having completed the description of a core, be it soil or rock, the final record to be included on the field log is the fracture state. The fracture state indices relate to the assessment of the mass state and are widely used in empirical correlations of the behaviour of the soil or rock mass. Empirical correlations with these indices include compressibility, blastability and ease of excavation, stand up times and support requirements in underground works.

Although the recording of these indices is often stated to be the last step in the logging process, this is not actually the case. Preparation for and early recording of information under this heading can usefully be started when first looking at the core and assessing what materials are present and how they are to be described. In particular it is a good idea at least to record the total core recoveries early in case there are core losses and gains, as discussed below.

TIP ON RECORDING FRACTURE STATE

- In the initial stages of preparing to log a core or other exposure, the logger should be tidying up the core or cleaning off the face. Whether this is to prepare for the core photography, or just to realign the core ready for logging or to expose undisturbed materials, the initial decisions on material and mass logging should be in preparation in the logger's mind.
- It is a useful aid at this stage to start to mark the natural fractures on the core box as shown in Figure 12.1. These marks should be depth correct and so will not be at the right inclination. When the logger is ready to measure the fracture state, this can be done using the marks on the box by measuring the relevant lengths down the centre marks on the wooden divider, which is equivalent to the core axis scan line. This is both more accurate and a whole lot easier than having to repeatedly pick up and return core lengths to the box separately for the measurement of each of the indices.
- This approach also has the advantage that any check logger coming along later can see the logger's thought process, which is very helpful to all parties.

In accordance with the overall rule that the logger is required to assess how the ground is in situ, measurements of the indices have to assess how the ground was before it was drilled. It is therefore a standard requirement in logging the fracture state that all artificial or induced fractures are excluded from the mass assessment. The fracture state record is an assessment requiring that all fractures which have been induced by the drilling or the creation of the exposure are ignored (see Chapter 6.1). The total core recovery is the only fracture index which is recorded as a fact in terms of the core that is in the box, not an interpreted condition in the ground.

It is easy to state that the induced fractures should be discounted in the logging, but it is often difficult to carry this out in practice. It takes a lot of practice, a good deal of common sense and the close examination of each and every fracture in a core box in order to make this assessment accurately and properly.

Figure 12.1 Core fractures being marked on the box (or tape). The mark for each fracture is depth correct, not inclination correct so as to allow measurement of spacings on the annotated record. Details of depth corrections, zones of non-intact core and core loss (AZCL) can also usefully be noted.

Table 12.1 Differences between natural and induced fractures

Observation	Characteristics of natural fracture	Characteristics of induced fracture
Minerals and grains	Appear matt, not shiny.	Appear bright, sparkly, fresh.
Staining of surface	Can be stained.	Generally not stained.
Surface form	Appearance is natural and surface is flat or parallel to other fractures.	Appearance is irregular and looks broken.
Fit of walls together	Good to poor	Generally reasonable to good
Polish and striation	Polish and striations if present tend to be linear and related to likely senses of movement.	Polish and striations if present tend to relate to direction of drilling and rotation of core bit.
Infill	Presence of mineral growth or weathering product infill	Infill can include evidence of cuttings and drilling flush.
Surface structures such as plume or hackle marks or conchoidal fractures	Structures are oversized with respect to core and not geometrically related to the core axis or circumference.	Structure size is compatible with core diameter and origin or boundary tends towards the core axis or circumference. Focal point of marks concords with impact points or splitting marks on core.

Features of a discontinuity surface that can be used to differentiate between natural and induced discontinuities include those given in Table 12.1. These criteria are derived from common practice and guidance provided by Kulander *et al.* (1990). If the distinction of induced from natural fractures is critical to the project, it may be best to consider some form of down hole imaging to record the fractures in situ.

There is an often stated convention in geotechnical practice that if the origin of a fracture is in doubt it should be assumed to be natural because this is conservative. However, this is only the case where the fractures are not helpful in the engineering context. Thus, for example, in assessing the rock stiffness for foundation settlement calculations, assuming all the doubtful fractures are natural puts more fractures into the model than are present in the ground, thereby reducing the mass modulus and maximising the calculated settlement. That is a conservative outcome and is good practice.

However, if the assessment is being made for the contractor who is going to excavate the foundation bases, this approach is non-conservative. In accordance with the convention, the mass assessment which includes all the doubtful fractures indicates the ground to be easier to dig than may be the case. An alternative approach can be followed in such cases which may help the excavating contractor. This alternative procedure might be to record the indices for all fractures present, on the basis that if the fracture opens up in drilling it is probably also available in the ground to help the bucket teeth or ripper point. This is a sensible possibility and just this sort of double assessment is carried out on a (very) limited number of projects. However, this approach presumes that the levels of disturbance imparted by the drilling and the excavator teeth are comparable. This is not usually something that can be pre-determined as correct and so there will always be questions about the actual relevance of the non-standard fracture indices in these instances. There is also concern that the care taken in the drilling and subsequent handling of cores is not a constant, so this approach can be fraught with uncertainty and can provide potentially misleading information.

It is also possible that the core is recovered in a significantly disturbed state where the outside surface is damaged or the diameter is reduced. As long as the internal portion of the core is considered to be representative of the in situ condition, a robust and practical view can be taken. The core in Figure 12.2 illustrates a damaged core with an outer skin of remoulded material and an 'inner' cut surface

(a)

Figure 12.2 Disturbed samples: (a) Damaged core that can still be logged despite the outer skin of remoulded material and the core catcher damage on the cut outer surface (photograph courtesy of Graham Elliott); (b) Driven thick walled core sample showing significant distortions to fabric.

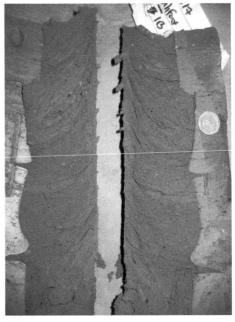

(b)

showing core cutting and catcher marks. This does not stop the heart of the core being logged as though the full intended diameter is present with no particular distinction being made in the recording of the indices outlined below.

TIP ON MEASURING FRACTURE INDICES
- It is suggested that the (very common) practice of calculating percentage ratios for the indices at the core box is unnecessary.
- The recoveries should be recorded and used as lengths in the field, as these are more meaningful than percentages.
- When the borehole log is typed up, the relevant computer program will ask for the recoveries as lengths and then calculate the percentages correctly.
- Therefore to carry out this calculation in the field is liable to error and is a waste of time.

In order to make a correct and realistic assessment of the fracture state for all parties who might be using the field logs on the project, it is clearly therefore imperative to recover the best quality cores possible and to log them carefully so that the recording of the fracture state can be correct and satisfy both of these opposing needs.

There are four records that are normally compiled in core logging in order to record the amount and quality of the core recovered, namely total core recovery (TCR), solid core recovery (SCR), rock quality designation (RQD) and fracture spacing or index (If or FI). The definitions from EN ISO 22475-1 and BS 5930: 1999 are given below. These fracture state records should be presented on the borehole log as part of the field record.

12.1 Total core recovery

The total core recovery (TCR) is the proportion of core recovered to the total length of core run. This includes both the solid (see definition below) and non-solid (or non-intact) core and is expressed on the field log as a percentage. Field measurements should be made and used in metres. The core run is the length reported by the driller as the depth interval for the coring trip.

One of the first steps that the logger should carry out is to assess any zones of core loss (AZCL). All zones of core loss should be assigned a depth. This information will need to be incorporated into the overall depth assessment of core runs and the depth of any logging observations. This is also necessary to ensure that the depths of samples are correct, which may require the records to be adjusted if core samples were extracted from the core box before logging is carried out and that the depths given on photographs are correct at the first time of asking. The assessed core losses will appear as percentages less than 100% in the TCR column, but additional information is usefully provided in the detail column about where in the core run the loss has arisen. If the core loss assessment is not made at an early stage, adjustments will need to be made subsequently in order to ensure a fully correct record. It is best practice to get the depths right first time by taking account of the losses in all depth related records as adjustments made later are time consuming and prone to error. The need to make depth adjustments later is an indication of a poor approach to systematic logging.

A not uncommon problem in addition to core loss is that of core gains, where core lost from one run is left down the hole but picked up on the subsequent core run. This will give core recovery of <100% in the first run, but may well result in recoveries >100% in the subsequent run. This needs to be addressed at the earliest possible stage of the logging process for all the reasons of ensuring correct depth records cited above. This is not however always possible, usually for reasons of complying with specification clauses relating to the time interval permitted before subsampling is carried out or that core resealing is to be completed. In these instances, the upper part of the lower core run will be moved in depth record terms to the base of the upper run. It is essential for the field logging record to indicate clearly the situation and to show the workings for correction of the depths. This may continue over a number of core runs until some fixed point is reached, such as the hole being dipped or completed.

Where core is dropped from one core run and then recovered at a second attempt in the next core run, this result can be identified by a note on the core recovered from the following run (CRF) being recorded in the detail description column. If the recommendations in this book are followed, the identification of core loss and subsequent recovery will be made before any other records are measured for depth. It is then a straightforward matter to mark the box with the corrected depths and all measurements will be correct at the first pass. It should be noted that the driller's core blocks must not be moved; these are primary records and are normally already included on the core photographs. If the early identification of core movements is not made, the correction of all measurements to the true depth is a tedious exercise. The adjustments need to be made throughout the core run, allowing for any zones of core loss and are applied to all measurements in the logging record as well as sample and test depths as outlined in Valentine and Norbury (2010).

Adjustment of the depths of core as a result of gains and losses will mean that the depth of subsamples may now conflict with the depths appended to the samples. In order to avoid corruption of the data set, the double depth records of each sample will need to be maintained and reconciled in an appropriate manner. The simplest solution is often to include these both on the field log in the format '*Sample number/Corrected depth/(Original depth)*'.

There is a further possible situation in measurement of TCR that is 'core swell'. Core of overconsolidated clays or mudrocks can in some circumstances swell so that measurements of TCR are consistently over 100% for a series of core runs. This is not usually more than about 5% but can locally increase to about 10%. This information should be reported as it occurs and does not require further comment on the field log. There will, however, be a potential problem in the TCR record on the field log as some log programs do not allow recoveries in excess of 100%. The field record should include a note of the actual TCR measured. To determine depth-related observations, it will be necessary to make any observations pro-rata through the core run length actually measured. This is unlikely to result in significant errors in depth reporting.

There is always a need to report the details of core loss wherever the loss is significant. What level of core loss is significant is always a problem. If you sat an exam and achieved a score of 90% or 95% you would be very proud, and rightly so. The situation is rather different in rock investigation as the 5% or 10% of core that is not recovered may be the portion that will control the rock mass behaviour. A core loss of say 10% in a 1.5 m run is 0.15 m, which is a significant length of missing information. This length of loss could arise for various reasons:

- Loss is due to damage in extracting the core from the core catcher box.
- Loss is due to a cavity or joint aperture(s) in the middle of the run.
- Loss is due to erosion of infill on the visible joints.
- Loss is due to core damage from cavings or stronger inclusions (e.g. flint or chert pebbles).
- Loss is due to the driller dropping a few pieces out of the core barrel.

These possible explanations differ widely in terms of the engineering behaviour of the ground. The significance of the most serious situations would normally be the cavity, open discontinuities or the erosion of infill which clearly would need to be correctly assessed. The assessment of the magnitude of these features will rely heavily on accurate measurement of the TCR.

The assessment of zones of core loss is usually not straightforward. Although core loss is identified, assigning it to a depth range in the core run may not be practical as there may be no particular evidence to guide that decision. In these circumstances a default position can be taken, with core losses being ascribed either to the tops or bottoms of all core runs which lack such evidence. It does not actually matter which option is used. This approach is one of minimising possible errors and, by its very arbitrariness, is one that is defensible.

As a general rule, loss of more than 5% should be recorded on the field log, but any loss exceeding 10% must be considered very carefully and reported accordingly with due attention being drawn to such considerations. This percentage criterion is not necessarily the same as that given in standard specifications where the driller is required to shorten core runs or take other improvement action when core losses occur. This situation has become more complicated with the introduction of semi-rigid core liners, which means that the driller does not see the core in order to make an accurate assessment of losses.

In some circumstances, core recoveries are much lower than the sort of figures mentioned above, for good reason. Not all ground is readily recoverable by core drilling. For instance, ground which includes materials with wide ranging strengths (flints in chalk, corestones in weathered granite, cobble and boulder soils) or rocks with very low strength (extremely weak mudstones, sandstones weathered to sand, worked ground) may not be recovered.

In these circumstances it is important that the field description achieves a balance between telling the reader about the ground at the borehole position (as required by BS 5930) and not overexposing the logger or their employer on liability. A scheme of wording in such cases is suggested in Table 12.2 to help loggers. Flexibility in the application of this suggested scheme will be necessary. Further discussion on this matter is given in Chapter 16.7 in connection with interpretation of in situ conditions when samples are disturbed or only partially recovered.

In addition to assessing the core losses, it is the responsibility of the logger to assess areas where core has been damaged due to drilling, as illustrated in Figure 12.2. No one else will be in a position to do this without coming back to open up the core boxes. The causes of core damage are often linked to losses of core and

Table 12.2 Scheme for recording low core recoveries

Indicative core recovery	Suggested approach	Descriptive format
75–100%	Record TCR, SCR, RQD and If as defined. Assess zones of core loss (AZCL) and assign depth ranges.	Carry out full description in accordance with standards. If core loss exceeds say 10% include statement such as 'Core loss presumed to be more weathered materials' or '…weaker materials…' or '…mudstone layers…' as appropriate.
50–75%	Record TCR, SCR, RQD and If as defined, i.e. percentages of full run length, not recovery. Assess zones of core loss (AZCL) and assign depth ranges if possible. Where this is not possible, emphasise loss and uncertainty about in situ conditions.	Suggested descriptive wording: 'Recovery is of stronger materials, weaker materials not recovered.' 'Recovered material is extremely weak low density white CHALK. Occasional flints. (Structured SEAFORD CHALK)'
25–50%	Record TCR, SCR and RQD as defined, i.e. percentages of full run length, not recovery. Record If where sensibly possible. The identification of the depth range of AZCL is unlikely to be possible.	Suggested descriptive wording: 'Partial recovery. Core loss presumed to be more weathered material.' 'Recovered core comprises medium strong grey coarse grained SANDSTONE. Sand and gravel size fragments recovered. Heavy discolouration on discontinuity surfaces, penetrating up to 3 mm. (probably weathered COAL MEASURES SANDSTONE)'
<25%	Record TCR Leave SCR, RQD, If columns all blank. Report lengths of core sticks recovered. The identification of the depth range of AZCL is almost certainly not possible.	Suggested descriptive wording: 'Minimal recovery. Core loss presumed to be more weathered material. Recovered core comprises GRAVEL and COBBLE size fragments of strong red coarse grained granite. (possibly weathered PETERHEAD GRANITE)'

resolving the evidence in the core box may well require discussion with the driller for any additional observations he or she may be able to contribute.

In any core where zones of core loss are associated with non-intact recovery and possibly also associated with strong inclusions such as pebbles or flint bands, for example, these aspects should all be considered together. As it is likely that the rock was at least largely intact in the ground, the logger is required to digest these elements of information for the benefit of all. The logger should stand back and take an overview on the interrelated elements of loss, disturbance and broken fragments and present this as a single detail comment, rather than as three separate comments.

EXAMPLE RECORDS OF CORE LOSS AND DAMAGE
- 8.2–8.8 m. Recovered core includes flint fragments and is non-intact. Assessed 200 mm of core loss. Flint band.
- 15.6–16.8 m. Core recovered non-intact with heavy staining on some surfaces and traces of clay infill. Probable vertical joint with very closely spaced horizontal joints.

12.2 Solid core recovery

The solid core recovery (SCR) is the proportion of solid core recovered to the total length of core run. The length should be recorded in the field in metres and allow the borehole logging program to calculate the reported expression as a percentage. The definition of solid core itself is of critical importance as the same definition is used in all the indices discussed below.

There are two definitions possible for solid core: the full diameter and the full circumference definitions. The full diameter definition should be used in practice.

Solid core is defined as core with at least one full diameter (but not necessarily a full circumference) measured along the core axis or other scan line between two natural fractures. The measurement of the length of solid core pieces forms a scan line survey. In principle, this scan line could be located anywhere on the core but, as core is not usually aligned within and between runs, it is usually safest to use the core axis as the scan line. This approach also has the advantage of making the measured solid core independent of core diameter.

It should be noted that in accordance with the full diameter definition, any core that has just a single set of inclined fractures would all be solid. Consider the core with a single set of parallel fractures shown in Figure 12.3. The central piece of core has a full diameter until the core axis daylights from the stick at either end, with the full diameter being drawn out of the paper. At the same point that one stick stops being full diameter for this reason, the next stick must start being solid core. Therefore only where there are two or more sets of fractures is the core not solid and for this to be the case, these fracture sets actually need to have different strike directions.

There have been various interpretations over the years of whether the full diameter or the full circumference definition is the right one. Those in favour of the full circumference definition included BS 5930: 1981, Hawkins (1986) and Mortimore (1990). Those in favour of the full diameter definition included EN ISO 22475-1, BS 5930: 1999, Norbury *et al.* (1986), Spink and Norbury (1990), Deere (1963), Deere and Miller (1966), Anon (1988) and ISRM (1981). The discussion seems to have settled on the full diameter definition.

The reason that the full diameter is the logical and correct definition to adopt is shown in Figure 12.4. In this diagram, the ground has a single set of fractures, shown inclined. In Core 1, drilled at right angles to this fracture set, all the core is solid and it makes no difference whether the full diameter or full circumference definition is followed as they give the same answer.

In Core 2 drilled at 45° to the set of fractures, the results from the two definitions differ. The shapes of the individual pieces of core are as in Figure 12.3 and so are all solid by the full diameter definition, but now only about a third of each stick actually has a full circumference. By the time we get to Core 3, drilled parallel to

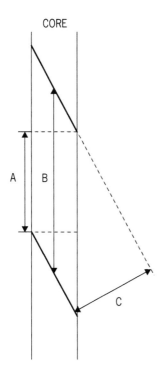

CORE

A | B

C

A Full circumference
B Full diameter
C Discontinuity spacing

Figure 12.3 Illustration of the various definitions of solid core which are available on a piece of core.

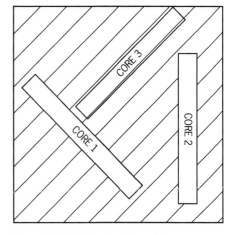

Figure 12.4 Illustration of the reason for the choice of solid core definition.

the fractures, there is no full circumference solid core but all the core remains full diameter by the solid definition.

Comparing the results from these three oriented cores, the measured fracture state indices are obtained as shown in Table 12.3.

It is clear that the full diameter definition gives more sensible results because regardless of the relative angle between the fractures and the borehole, the ground is reported to be the same quality overall. If the full circumference definition is

Table 12.3 Comparison of results from different definitions

Index	Core 1	Core 2	Core 3
TCR	Full circumference = 100%	Full circumference = 33%	Full circumference = 0%
	Full diameter = 100%	Full diameter = 100%	Full diameter = 100%
SCR	Full circumference = 100%	Full circumference = 33%	Full circumference = 0%
	Full diameter = 100%	Full diameter = 100%	Full diameter = 100%
RQD	Full circumference = 100%	Full circumference = 33%	Full circumference = 0%
	Full diameter = 100%	Full diameter = 100%	Full diameter = 100%

used, the ground quality reported in this case reduces to zero, yet the actual ground has not changed at all. Although it is true that the inclination of discontinuities in relation to the cores or, more importantly, the engineering works such as the slope or underground opening is also of critical importance, these indices are recording the rock mass quality in a general sense for comparison with other areas on the site or for correlation with precedent experience.

Several years ago certain investigations in the southwest of England were using the full circumference definition for all core logging and, as a result, all the rocks around granite intrusions like Dartmoor, which have steeply dipping bedding or cleavage, had SCR = RQD = 0% and an If of non-intact. The construction of road cuttings through these materials required blasting, which is hardly consistent with the reported index values. A more sensible and useful outcome would have been obtained using the full diameter definition.

12.3 Rock quality designation

The rock quality designation (RQD) is the total length of all the solid core pieces each of which is greater than 100 mm in length between natural fractures. Within each core run, the lengths of these individual pieces are summed to give a total length for the run in metres. This index is also expressed as a percentage of the total length of the core run. The RQD is conventionally assessed for each core run.

Classification terms are provided for use in a summary text to describe the quality of the mass based on RQD. These terms are presented in Table 12.4, but are not used in practice on field logs.

Table 12.4 Descriptive terms relating RQD to rock quality

RQD (%)	Term
0 to 25	Very poor
25 to 50	Poor
50 to 75	Fair
75 to 90	Good
90 to 100	Excellent

RQD is used as a standard parameter in records of core logging. The index was originally introduced for use with NX size core (54.7-mm diameter) and the dimension of 100 mm provided an indication of the proportion of the core run that might be suitable for laboratory testing although this size requirement is no longer necessary. RQD has considerable value in estimating the quality of the rock mass and a number of empirical applications have been derived. These include the j factor to take account of fracturing in adjusting rock material stiffness to that of the overall rock mass and as an input to assessments of the ease of excavation of rock in surface excavation. RQD is particularly widely used in the assessment of stability and support of rock tunnels. The index has been used as a flag to identify low quality rock zones and forms a basic element value of the major mass classification systems such as the rock mass rating (RMR) and Q-systems.

Although RQD is a simple and inexpensive index, it cannot characterise the rock mass alone as no details are included on the orientation or characteristics of the discontinuities, see Chapter 11.

12.4 Fracture spacing

The fracture spacing is the average length of solid core pieces between natural fractures. The spacing is a scan line (see above) measured over core lengths of reasonably uniform characteristics, not core runs. The fracture spacing is usually measured in millimetres and is usefully given by estimating the minimum, mode and maximum fracture spacings in millimetres for any uniform length, for example 100/250/300. The core pieces that represent the minimum and maximum If measurements will actually exist in the core boxes as the shortest and longest sticks present. The modal If will be the typical stick length and may not actually exist as a piece in the recovered core.

'Non-intact' (NI) is the term used when the core is fragmented and indicates that the core is not solid.

Some companies use fracture index (FI) which is the inverse of fracture spacing, that is fractures per metre. Fracture spacing has the advantage of being a direct measurement that is easily envisaged, whereas the inverse does not provide such an immediate picture to the reader.

The recording of fracture spacing or fracture index zones should not be made over the same lengths as the core runs, although many companies do follow this practice. This loses valuable information about changes and boundaries in the strata. A common example is that the If units can usefully match the strata in an interbedded sequence such as one of mudstone and sandstone and this would provide a more relevant and comprehensive record of the fracture state in each separate lithology.

The application of these indices and their definitions are illustrated in Figure 12.5. In this diagram it can be clearly seen that the total core includes not only all the solid core section on the left, but also the non-intact section towards the right. The only core that is non-total is the core loss at the right hand end of the core run.

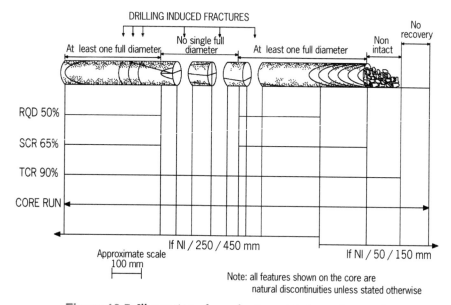

Figure 12.5 Illustration of core fracture state terms (see text).

The solid core includes all that is TCR, but not the section of broken fragments to the right. In addition, the section in the middle, where the core run has been pulled apart for illustration, shows that the core is not solid because a full diameter is not present owing to the intersection of two fractures with a different strike.

Similarly all of the core that is solid could count towards the RQD, but the natural parallel fractures to the right of centre are less than 100 mm apart and so this does not count. No section that is not solid can count towards RQD.

Finally the fracture spacing can be assessed without linkage to the core runs. A boundary is selected reflecting the change in fracture spacing and continuing to the next core which runs in both directions.

The definitions given above are those that are used widely internationally. However, these definitions are arbitrary and alternative definitions are used on occasion. For example, some alternatives that have been suggested are given below:

- Measurement of indices takes account of all the fractures present in the core rather than just those assessed as being natural fractures (see above).
- Lithology quality designation (LQD): Hawkins (1986) advocated measurement of RQD values by lithological unit rather than core runs, in order to reflect better the quality of the respective lithologies, much as is suggested above in the context of fracture spacing measurements.
- RQD can be recorded for any length of pieces other than the 100 mm defined as the standard. Any length could be used and Hawkins (1986) suggested RQD_{300} as a possibility. It should be noted that all the empirical correlations using RQD (rock mass stiffness, rock mass ratings) use RQD_{100}.

Note that it is crucial that if any non-standard approach is used in recording any of the indices measuring the fracture state, this must be stated on the borehole logs. It can be stated in the report text as well, but it is often the case that text and field logs tend to become separated in time.

Rotary coring is often carried out nowadays in a number of formations other than rock, including old foundations or brickwork and in soils. A slightly modified approach needs to be adopted in accordance with common sense since there is little benefit in recording the fracture spacing in a clay or brickwork.

If coring in soils, TCR is the only index that should be recorded, in accordance with Clause 44.4.4, BS 5930: 1999. This rule also needs to be followed when

TIP ON MEASURING CORE INDICES
- While measuring the TCR, SCR and RQD, loggers should be measuring the total recovery first, as required in any case to obtain all the correct depth-related records in the first pass.
- In making the subsequent measurements of SCR and RQD, it is usually easier and quicker to identify those lengths of core that do NOT qualify for the latter indices and to subtract these from the TCR in turn. This is easier than remeasuring the whole core run three times, which is what most inexperienced loggers do.
- Marking all the natural fractures on the core box in the first instance is also a really helpful practice (see Figure 12.1), as the SCR, RQD and If measurements can then be made directly from the marks without having to repeatedly lift the core pieces from the box.
- This system will result in less mental effort and thus should also result in better and more accurate recording of the indices.

recording the condition of cores at the soil/rock interface and requires common sense. For example, a core run which is 50% very stiff clay and 50% extremely weak mudstone can have up to 100% TCR, but only up to 50% SCR or RQD, although obviously it is unlikely that the rock section of the core will be of such high quality in practice.

When coring is carried out in concrete or brickwork core, the indices are of no benefit. After all what is the definition of a natural compared with an induced fracture in these circumstances and what is the benefit of such information? The answer to this question is clear. It is, however, good practice to record TCR in all cores. There might be occasions when a record of SCR could be useful in concrete, rather than brickwork, in which case it can be recorded where useful. RQD and If are not sensibly applicable in practice and so recording of these indices should not be carried out.

13 Low Density Soils

There are a number of soils and rocks that are characterised by an unusually low density. These materials are often tricky in an engineering sense, not least because their low density tends to reduce their strength and make them sensitive to disturbance. This sensitivity applies to both the proposed engineering works and to the sampling as part of the investigation. It is therefore appropriate to give these materials special care and attention when contemplating their description for engineering purposes.

On the flow chart for soil description presented in EN ISO 14688-1 (see Figure 4.1) a number of low density soils are referred to. The first are the organic soils and the second are soils that have low density, but are inorganic. This class is indicated in EN ISO 14688-1 to comprise only volcanic soils. There are, however, other low density inorganic soils such as wind blown deposits. The most common example of this type of soil is loess.

The soils and rocks that will be covered in this section are the organic soils which include topsoil and peat as well as transported organic materials, volcanic soils and rocks and loess or other wind blown materials.

13.1 Organic soils

13.1.1 Topsoil

Topsoil and the underlying subsoil are the most commonly encountered organic soil strata. Thirty years ago almost every hole started in topsoil, but with the advent of more brownfield sites, this is less commonly the case nowadays. To most engineers, topsoil is the material that is stripped from the site as the first action in site preparation, stored and returned in some form at the end of the works. On this basis, most engineers do not want to know anything about the topsoil. This attitude forms the background so that investigations generally pay only limited attention to topsoil. Topsoil is not generally sampled and so is not logged by the geologist but instead reliance is placed on the 'driller's description' which is usually just the word 'topsoil'. Indeed the topsoil is usually not accurately recorded in terms of either its character or its thickness.

Part of the reason for these inaccuracies is that topsoil is variously defined. The minimal definition is that the topsoil is the layer of turf which comprises the grass and its mat of rootlets. Others may take it as the turf plus the brown or dark brown organic rich layer beneath which also contains abundant rootlets; this is the layer that supports plant life. The maximal definition is that of the soil scientists who include the additional subsoil horizons which together make up the whole of the profile that is the soil and the surface activities it supports.

The approach to investigation, sampling and testing of topsoil in order to provide the sort of information required for a detailed assessment of the whole soil profile for re-use is given in BS 3882: 2007 and ISO 25177: 2008.

The BS standard provides a classification and testing scheme but does not include a descriptive system and so is not used in producing field logs. The ISO standard does provide a descriptive scheme which is based on the World Reference Base for Soil Resources (FAO, 1998). This descriptive scheme is extensive and elaborate and provides substantial detail on soil and soil profiles. This scheme is too detailed and wide ranging to be routinely used in engineering practice so that descriptions compiled following this approach are not expected to appear on a site investigation field log. For this reason the FAO scheme is not presented here, but the full scheme can be downloaded from www.fao.org/nr/land/soils/soil/en.

A classification based on soil constituents is given as shown in Figure 13.1. There are a number of other examples given in ISO 25177.

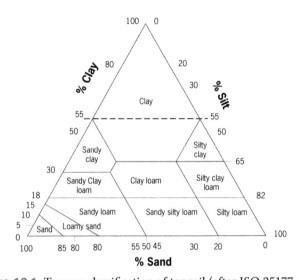

Figure 13.1 Ternary classification of topsoil (after ISO 25177: 2008).

Logging the full profile to the level of detail incorporated by soil scientists is a rare event in practice. However, such detail can provide information pertinent to the separation and storage of the different layers for optimum site restoration. However, in engineering geological investigations the normal option is to take the middle of the three definitions above, namely to consider the topsoil as the turf plus the underlying brown or dark brown organic rich layer with abundant rootlets. Just recording the turf will miss out any information on the organic rich layer beneath the turf, which may be useful for the site restoration activity and may also be compressible and weaker than the underlying 'engineering' soils.

The description of topsoil itself is not covered in any of the Standards routinely used in relation to site investigation. It is therefore recommended that topsoil should be described in accordance with the systematic approach outlined below.

BACKGROUND ON THE TERM LOAM

- Soils are described by soil scientists for environmental purposes as clay soil, sand soil or loam soil (see Figure 13.1).
- Loam soils form a significant majority of agricultural or top soils, generally reckoned to be over 75%.
- There are no common definitions of loam, although it can be taken to be a soil composed of sand, silt and clay in relatively even concentration (about 40–40–20% concentration, respectively), considered ideal for gardening and agricultural uses. Loam soils generally contain more nutrients and humus than sandy soils, have better infiltration and drainage than silty soils and are easier to till than clay soils.
- In describing soils for engineering purposes in site investigations, it is suggested that topsoil should be described in accordance with the guidance in this section. These descriptions should be included on field logs.
- The use of any other descriptive approach to the description and classification of soils which is used should be recorded separately, in a separate log or schedule.

A simple word order is adopted. This is no different from the general word order used in all descriptions, it is just that it is usually appropriate to record slightly different information. The word order for topsoil is:

- Consistency using the hand tests defined in Chapter 5.2 if an appropriate exposure is available;
- Colour using the approach outlined in Chapter 7;
- Secondary constituents using the terms and approach outlined in Chapter 8;
- The material type should be a principal soil type assessed following the approach and terms given in Chapter 4. The practice of merely describing the material as 'topsoil' is not favoured, the use of a principal soil term being preferred with the identification of 'topsoil' being left to be the geological formation as in the second example below;
- The presence and size of roots and rootlets should be described using the terms indicated in Table 13.1.

EXAMPLE DESCRIPTIONS

- (Friable brown clayey TOPSOIL with occasional gravel. Matted rootlets) – approach not preferred
- Firm dark brown sandy CLAY with common rootlets and rare roots up to 20 mm diameter (TOPSOIL)
- Dark brown sandy CLAY with few rootlet holes (RELICT TOPSOIL)

13.1.2 Peat

Peats are materials that accumulate in situ in a mire or bog. The peat consists of organic material derived from plants which is in various stages of decomposition and can include significant lateral and vertical variability owing to changes in depositional conditions, including changes in ambient temperature. The stage of decomposition can be remarkably low as the materials are preserved under water in an

Table 13.1 Definition of roots and rootlets

Roots are ubiquitous in top soil and have a major effect on the properties of the soil, in either a soil science or an engineering context. It is therefore important to record their dimensions and to give some indication of their frequency/spacing/abundance.

Size
The following definitions of root sizes is given in ISO 25177:
- Very fine roots   <0.5 mm diameter
- Fine roots   0.5–2.0 mm diameter
- Medium roots   2.0–5 mm diameter
- Coarse roots   >5 mm diameter

- It is also important to note if these features are represented by root holes rather than fibres as these are high permeability conduits.

Spacing
If detail is required (which is probably a rare occurrence), the roots can be given spacings using the terms for fracture spacing. However as the minimum category is 20 mm, this is rather a crude scale. The spacing can either be given as measured in millimetres or combined into the frequency term as indicated below.

Frequency
The simplest way to describe the frequency of roots is normally to use the terms provided for description of minor constituents so examples could be 'with rare'/'with occasional'/'with frequent'/'with abundant'. A true turf layer may well actually have so many rootlets that they are matted – a good term to use. The advantage of these terms is that they can be used in a qualitative sense which is appropriate in this context.

The definitions of abundance given in ISO 25177 are as follows:
- Very few   1–20 roots per dm^2, that is a 10 × 10 cm^2
- Few   20–50 roots per dm^2
- Common   50–200 roots per dm^2
- Many   >200 roots per dm^2.

These definitions may prove useful on occasion but will be difficult if the roots are smaller than medium in size and are not recommended for general use in site engineering geological description.

---

anoxic environment. Peats are usually dark brown or black in colour as shown in Figure 13.2, have a distinctive smell and low bulk density (about 1.1 Mg m^{-3} compared with a normal inorganic soil with a density of about 2.0 Mg m^{-3}) and extremely high water contents of up to 1000%. Peats can also include inorganic particles which will be either disseminated throughout the peat or occur as discrete inclusions.

Peats pose significant engineering problems through their high compressibility, creep movements, low strength and possibility for gas generation. A full description of the morphology and character of peats is given in Hobbs (1986).

Peat that is completely free of extraneous mineral matter may have an ash content as low as 2%, that is an organic content exceeding 98%. There is a variety of schemes in use for the description and classification of peats and related organic soils which are based on assessment of the constituents of the peat together and its state of decomposition. The decomposition of the peat represents the breakdown of the structure of the organic materials and their transformation into amorphous material.

Figure 13.2 Typical peat showing both decomposed material and fibres.

The approach to the description of peats is covered in Codes and Standards only superficially and with a basic descriptive approach provided. Classification is given little more than a passing mention. Most of the detail is only given by reference to other guidance. The most recent peat naming terms given in EN ISO 14688-1 include two forms of peat which are made up of plant remains together with animal excreta and remains (see Table 13.4).

Because of the variety of approaches that are available, the approach to be followed on any particular project will need to be established in advance. This will then need to be included in the contract documents to allow logging contractors to assess correctly the rate at which logging can be carried out. If the approach to be followed is not defined, the default procedure will be to describe in accordance with BS 5930: 1999 as shown below. Use of this approach indicates that more detailed description and classification are not required and so is an unnecessary effort.

The guidance on detailed description and classification that is most commonly used in practice is that given in the Scottish Executive Guidance (Anon, 2006, Section 4.4.5) and which itself includes two schemes:

- modified von Post (MVP) classification, as outlined in Hobbs (1986)
- Troels-Smith (TS) classification, as outlined in Long *et al.* (1999).

The choice of which of these various approaches might be appropriate and which tends to be used in actual practice is summarised in Table 13.2. Details of the three approaches are then outlined below.

Materials other than peat, such as organic transported soils should be described in accordance with Section13.2.

13.1.2.1 The BS 5930 descriptive scheme

This scheme is used when peat is neither extensive nor an undue problem and thus where the basic engineering solution would be a simple dig out and replace. It is therefore appropriate to have available this straightforward scheme which records the presence and thickness of compressible organic soils and provides some

Table 13.2 Selection of approach to follow in peat description

	BS 5930: 1999	MVP – written format. Add shorthand notation only when explicitly required to do so	TS – shorthand notation only
Simple – where the basic engineering solution would be a simple dig out and replace.	Yes	No	No
Complex in England – where the peat is significant in engineering terms or when called for by the contract documents.	No	Yes	No
Complex under Scottish Executive documents – when called for by the contract documents.	No	Yes	Yes

information on the character of the soil. If more detail is required, because the material is to be engineered in some way, one of the more detailed approaches given below should be used.

Peat is identified and described according to its condition and type. The standard word order to be used is:

- condition (this is a change from earlier practice where this was termed consistency, but has been changed to avoid clash with the consistency as described in inorganic fine soils). Terms are given in Table 13.3.
- colour
- secondary constituent
- type
- peat.

The colour of peat is usually black although brown can occur. It should be noted that the properties of peat can change rapidly on exposure (e.g. colour and moisture content). Loggers should be aware of this and complete descriptions should record any such changes.

Secondary constituents such as sand or organic materials can be incorporated into the organic soil. The organic materials can include wood or animal remains or other foreign items. Peats may be described, for example, as slightly clayey or very

Table 13.3 Terms for description of peat condition

Term	Definition
Firm	Fibres compressed together
Spongy	Very compressible; open structure
Plastic	Can be moulded in hand. Smears fingers with soapy feel.

Table 13.4 Types of peat

Type	Field assessment
Fibrous	Plant remains recognisable and retain some strength.
	Water and no solids exude between fingers on squeezing.
Pseudo-fibrous	Plant remains recognisable and strength lost.
	Turbid water and <50% solids exude between fingers on squeezing.
Amorphous	No recognisable plant remains; mushy consistency.
	Paste and >50% solids exude between fingers on squeezing.
Gyttja	Decomposed plant and animal remains.
	May contain inorganic particles.
Humus	Remains of plants, organisms and excretions with inorganic particles.

sandy. These adjectival descriptors are used qualitatively here, as distinct from the defined terminology in inorganic soils. It is important to distinguish between disseminated or discrete secondary components. Thus a 'sandy PEAT' is not the same as a 'PEAT with frequent pockets of sand'.

The type of peat is determined in the wet state by inspection and by squeezing in the hand in order to assess its fibre content. The terms in Table 13.4 should be used and some example descriptions are provided below. The traditional description of peats has generally been using the first three terms in Table 13.4 which considers peats composed purely of plant remains. The inclusion of animal remains or excretions introduces the need for the additional terms in the last two rows. It should be noted that these materials can also be present in sewage sludge or night soil deposits, but these should be described as made ground.

EXAMPLE DESCRIPTIONS

- Firm black oxidising to brown fibrous PEAT
- Plastic dark brown slightly sandy amorphous PEAT with occasional small pockets up to 5 mm of grey clay
- Spongy brown HUMUS with occasional gravel size fragments of wood.

Identification of the degree of humification is often a useful adjunct to the description. The notation for this description (e.g. H5) is given within the von Post scheme presented below.

13.1.2.2 Detailed description and classification schemes

Peat materials in trial pits and borehole samples can be described and classified using both the modified von Post (MVP) and Troels-Smith (TS) schemes where they are significant in engineering terms or when appropriate for the project, see Table 13.2. The schemes are usually presented together on the same log which might require some modification of the field log format.

When the MVP and TS schemes are used, it is not necessary also to continue to use the straightforward BS 5930 approach.

The MVP and TS schemes each include a written format and a shorthand notation. There are advantages to both approaches; the written style is more readily understood, checked and read but once a logger has got used to the shorthand terms, these provide a rapid and convenient approach for use on sample description sheets.

As is normal practice in compiling field logs, the logger should identify a small number of main strata and use the 'detail' entry to describe thin bands or minor changes within main strata.

13.1.2.3 Modified von Post scheme, after Hobbs (1986)

General
The modified (or extended) scheme is described in Appendix A of Hobbs (1986) and is presented below. Use of the modified von Post (MVP) scheme requires reference to Appendix A of Hobbs (op cit) which should be read and used in conjunction with the qualifications and amendments added within the text below.

Peats are composed of the partly decomposed remains of plant communities containing varying morphology and texture. It is this structure that affects the retention or expulsion of water in the system, gives it its strength and ultimately differentiates one type of peat from another. The von Post classification attempts to describe peat and its structure in quantitative terms and the extended system set out below provides a means of correlating the types of peat with their physical, chemical and structural properties.

The full field description covers colour, designation of plant types, humification (H), water content (B), fine fibres (F), coarse fibres (R), wood remnants (W), shrub remnants (N), tensile strength (TV and TH), smell (A), plasticity (P) and acidity (pH). Classification of organic content (N) requires laboratory testing and so is not included in the field description.

Colour
Colour and any changes such as oxidation effects are to be described.

Plant types

- Bryales (moss)   = B
- Carex (sedge)   = C
- Equisetum (horse tail)   = Eq
- Eriophorum (cotton-grass)   = Er
- Hypnum (moss)   = H
- Lignidi (wood)   = W
- Nanolignidi (shrubs)   = N
- Phragmites   = Ph
- Scheuchzeria (aquatic herbs)   = Sch
- Sphagnum (moss)   = S

Designation
With few exceptions natural peats consist of a mixture of two or more plant types. The designation adopted is to list the plant types in descending order of content, that is, the first symbol represents the principal component. For example, a peat classified as ErCS consists mainly of Eriophorum remnants, while the content of Carex remnants would be lower and that of Sphagnum remnants relatively low. The designation is omitted when plant types cannot be identified.

Designation of plant types would often require the services of or guidance from a specialist botanist and therefore may well not be part of the field description.

Humification (H)

The degree of humification is graded on a scale from H1 to H10. The various degrees of humification are recognized by the hand tests given in Table 13.5 below.

As peat is not generally uniformly decomposed, the humification identification number for a stratum should be an average based on several squeezing tests. The

Table 13.5 Degrees of humification

Degree of humification	Decomposition	Plant structure	Content of amorphous material	Material extruded on squeezing (passing between fingers)	Nature of residue
H1	None	Easily identified	None	Clear, colourless water	
H2	Insignificant	Easily identified	None	Yellowish water	
H3	Very slight	Still identifiable	Slight	Brown, muddy water no peat	Not pasty
H4	Slight	Not easily identified	Some	Dark brown, muddy water; no peat	Somewhat pasty
H5	Moderate	Recognisable, but vague	Considerable	Muddy water and some peat	Strongly pasty
H6	Moderately strong	Indistinct (more distinct after squeezing)	Considerable	About one-third of peat squeezed out; water dark brown	Fibres and roots more resistant to decomposition
H7	Strong	Faintly recognisable	High	About one-half of peat squeezed out; any water very dark brown	
H8	Very strong	Very indistinct	High	About two-thirds of peat squeezed out; also some pasty water	
H9	Nearly complete	Almost unrecognisable		Nearly all the peat squeezed out as a fairly uniform paste	
H10	Complete	Not discernible		All the peat passes between the fingers no free water visible	

squeeze test is not applicable to compressed or friable peats, the equivalent humification number for which should be obtained by close examination. The squeeze test may not be applicable to peats in which clay can be clearly seen.

Loggers should be aware that some peats are pseudo-fibrous which have an apparent fibre structure but are really amorphous. Conversely some fine fibrous peat may appear granular which implies a high degree of humification.

Water content (B)

In the field the water content of the peat is estimated on a scale from B1 (dry) to B5 (very high). In terms of actual water contents the following ranges are suggested:

- B1 dry;
- B2 less than 500%;
- B3 500 to 1000%;
- B4 1000 to 2000%; and
- B5 greater than 2000%.

It is known that water content tends to decrease with humification and may be roughly correlated with broad peat categories. However the information provided does not offer guidance on estimation of water content in the field and further guidance would be required for this to be carried out. In the absence of guidance, classification may be based on the results of laboratory tests and will not be part of the field description.

Fine fibres (F)

Fine fibres are defined as fibres and stems smaller than 1 mm in diameter or width (in contrast to Table 13.1). They are often of the Eriophorum species, but Hypnum or Sphagnum stems may also be included if properly specified, e.g. F(H) or F(S). Shrub rootlets may also be included, specified as F(N). No special designation is indicated for plant root hairs, rhizoids, or other fibres. The content of fine fibres is graded on a scale from 0 to 3 as follows:

- F0 nil;
- F1 low content;
- F2 moderate content; and
- F3 high content.

Coarse fibres (R)

Coarse fibres are defined as fibres, stems and rootlets greater than 1 mm in diameter or width. They are often of the Carex genus, but Hypnum and Sphagnum stems may also be included if properly specified, for example R(H) or R(S). Shrub (N) rootlets are specified as R(N). The content of coarse fibres is graded on a scale from 0 to 3 as follows:

- R0 nil;
- R1 low content;
- R2 moderate content; and
- R3 high content.

The proportion of fine (F) and coarse (C) fibres should be assessed on a scale of 0 being nil to 3 being high. The proportions of fine and coarse fibres relate to the proportion of each within the fibrous content and thus should sum to no more than 100%. The proportion of fibres within the whole peat is taken together with the amorphous content and indicated using the humification scale H. In this context pseudo-fibres are included with the amorphous content.

Wood (W) and shrub (N) remnants

Wood and shrub content is graded on a scale from 0 to 3 as follows:

- W0 nil;
- W1 low content;
- W2 moderate content; and
- W3 high content.

The original designation for macroscopic wood (W) and shrub (N) remnants was, collectively, V (from Swedish vedrester = wood remnants; von Post, 1922).

Unlike the fibres, the proportions of wood (W) and shrub (N) remnants, given by similar scales of 0 to 3 are taken to be the proportion of each within the whole peat in relative terms.

Additional information

The above system, which has been taken from Landva and Pheeney (1980), can easily be extended by the addition of symbols for organic content, structural aniso-tropy, smell, plasticity and acidity, if required.

Organic content (N)

It is not possible to estimate the organic content unless the peat is obviously clayey when the von Post humification test would not be realistic. Following ignition loss determinations the organic content may be graded as follows:

- N5 greater than 95% organic matter;
- N4 95–80%;
- N3 80–60%;
- N2 60–40%; and
- N1 40–20%.

Tensile strength (TV and TH)

The tensile strength in the vertical and horizontal directions may be judged by pull-ing specimens apart in these directions. The following scale may be used:

- T0 zero strength;
- T1 low, say less than 2 kN/m^2;
- T2 moderate, say 2–10 kN/m^2; and
- T3 high or greater than 10 kN/m^2.

It is known that the amorphous peats tend to have lower strength and fibrous peats to have higher strength. However the information provided does not offer guid-ance on estimation of tensile strength and further information would be required for this to be carried out. In the meantime this should not be part of the field description.

Smell (A)

The smell, which is an indication of fermentation under anaerobic conditions, may be scaled as follows:

- A0 no smell;
- A1 slight;
- A2 moderate;
- A3 strong.

Methane, CH_4, the main indicator of anaerobic activity has no smell. If it is specifically detected this should be reported.

Plasticity (P)

Plastic limit test possible P1; not possible P0.

In general plasticity tests can only be carried out on peats containing clay. The plastic index test should be attempted, the water content being adjusted as necessary by addition of water to dry material or by working wet material in the hand.

Acidity (pH)

- Acid pH_L;
- neutral pH_0;
- alkaline pH_H.

The pH of peat water can be simply measured in the field with litmus paper.

Proposed classification – word order

Logging can be carried out using the written format and/or the shorthand notation. The written descriptions should also include colour and the humification classifier.

The standard sequence for descriptions (written or shorthand) is colour, humification, fine fibres, coarse fibres, wood and shrub remnants, smell, plasticity, acidity.

Examples of the application of the von Post scheme are given below.

EXAMPLE DESCRIPTIONS

The basic level of MVP is:

- Dark brown oxidising to black moderately decomposed H5 high fine fibrous PEAT with low coarse fibres and some amorphous material. No smell, plastic limit test possible, acid.
- The equivalent shorthand notation is H5 F2 R1 W0 A0 P1 pH1.

  Use of the extended MVP with ignition loss and pH determinations is:

- Dark brown, oxidising to black, moderately decomposed H5, mainly fine fibrous PEAT with some coarse fibres and amorphous material. Low vertical tensile strength, moderate horizontally. No smell, plastic limit test possible, neutral. Genera not identified.
- The equivalent shorthand notation is H5 B2 F3 R1 W0 N3 TV1 TH2 A0 P1 pH0.

13.1.2.4 Troels-Smith scheme

Details of the Troels-Smith system are given on pages 269 to 274 of Long *et al.* (1999). The scheme covers all soils but is only generally applied to peats. The subjective and qualitative scheme is used to identify the components and physical properties of each main stratum using abbreviated codes. There is a written scheme of

Table 13.6 A selection of the commonest sediment types recorded in coastal sediments (after Troels-Smith, 1955)

Name	Code	Sediment type	Field characteristics
Argilla steatodes	As	Clay <0.002 mm	May be rolled into a thread ≤2 mm diameter without breaking. Plastic when wet, hard when dry.
Argilla granosa	Ag	Silt 0.06–0.002 mm	Will not roll into thread without splitting. Will rub into dust on drying (such as on hands). Gritty on back of teeth.
Grana minora	Gmin	Fine, medium and coarse sand (0.06–2.0 mm)	Crunchy between teeth. Lacks cohesion when dry. Grains visible to naked eye.
Grana majora	Gmaj	Fine, medium and coarse gravel (2–60 mm)	
Testae (molluscorum)	test. (moll)	Whole mollusc shells	
Particulae testarum (molluscorum)	part test. (moll)	Shell fragments	
Substantia humosa	Sh	Humified organics beyond identification	Fully disintegrated deposit lacking macroscopic structure, usually dark brown or black.
Turfa herbacea	Th	Roots, stems and rhizomes of herbaceous plants	Can be seen vertically aligned or matted within sediment in growth position.
Turfa bryophytica	Tb	The protonema, rhizods, stems, leaves etc. of mosses	Can be seen vertically aligned or matted sediment in growth position.
Turfa lignosa	Tl	Roots and stumps of woody plants and their trunks, branches and twigs	Can be seen vertically aligned or layered within sediment in growth position.
Detritus lignosus	Dl	Detrital fragments of wood and bark >2 mm	Non-vertical or random alignment, may be laminated, not in growth position.
Detritus herbosus	Dh	Fragments of stems and leaves of herbaceous plants >2 mm	Non-vertical or random alignment, may be laminated, not in growth position.
Detritus granosus	Dg	Woody and herbaceous humified plant remains 0.1–2 mm that cannot be separated	Non-vertical or random alignment, may be laminated, not in growth position.
Limus detrituosus	Ld	Fine detritus organic mud (particles <0.1 mm)	Homogeneous, non-plastic, often becomes darker on oxidation and will shrink on drying. Most shades of colour.
Limus ferrugineus	Lf	Mineral and/or organic iron oxide	Forms mottled staining. Can be crushed between fingers. Often occurs in root channels or surrounding Th
Anthrax	Anth	Charcoal	Crunchy black fragments.
Stirpes	Stirp	Tree stump	
Sratum confusum	Sc	Disturbed stratum	

descriptors, but this applies to individual components only, is not straightforward and its adoption is not recommended. Information additional to the shorthand can be provided where relevant, such as descriptors of colour and bedding.

Codes for the proportions of components are given in Table 13.6 and codes for physical properties in Figure 13.3. These cover the degree of darkness, stratification, degree of dryness, degree of elasticity and boundary contact with the overlying stratum. Codes for the degree of humification (humicity) are given in Table 13.7. These tables and the figure are presented for reference. Identification

Nigror			(Degree of darkness)
		0	The shade of quartz sand
		1	The shade of calcareous clay
		2	The shade of grey clay
		3	The shade of partly decomposed peat
		4	The shade of quartz black, fully decomposed peat
Stratification			(Degree of stratification)
		0	Complete heterogeneity; breaks equally in all directions
		1	Intermediate between 0 and 4
		2	Intermediate between 0 and 4
		3	Intermediate between 0 and 4
		4	Very thin horizontal layers that split horizontally
Siccitas			(Degree of dryness)
		0	Clean water
		1	Thoroughly saturated, very wet
		2	Saturated
		3	Not saturated
		4	Air dry
Elasticitas			(Degree of elasticity)
		0	Totally inelastic, plastic
		1	Intermediate between 0 and 4
		2	Intermediate between 0 and 4
		3	Intermediate between 0 and 4
		4	Elastic
Limes superior			(Boundary)
		0	>1 cm boundary area - *diffusus*
		1	<1 cm and >2 mm - *conspicuus*
		2	<2 mm and >1 mm - *manifestus*
		3	<1 mm and > 0.5 mm - *acutus*
		4	<0.5 mm

Figure 13.3 Physical properties of a sediment as well as boundary classifications (after Troels-Smith, 1955). The symbols representative of each element are also shown.

Table 13.7 Terms for the assessment of humicity (after Troels-Smith, 1955)

Humicity	Comment
0	Plant structure fresh. Yields colourless water on squeezing.
1	Plant structure well-preserved. Squeezing yields dark coloured water. 25% deposit squeezes through fingers.
2	Plant structure partially decayed though distinct. Squeezing yields 50% deposit through fingers.
3	Plant structure decayed and indistinct. Squeezing yields 75% deposit through fingers.
4	Plant structure barely discernable or absent. 100% passes through fingers on squeezing.

of components of peat requires various types and sizes of plant remains of herbaceous plants, mosses and woody remains to be identified. This can be carried out at a basic level by loggers who are not botanists provided some specific training is first provided.

The components of a layer are recorded on a four point scale where 1 is 25% and 4 is 100% of the layer. Thus a peat which comprises 100% herbaceous roots with attached stems and leaves (turfa) would be described as Th4, whereas a peat with 50% clay and 50% herbaceous stems and roots as As2, Th2. Additional minor components (<25%) which do not contribute to the main component totals can be recorded by a '+'. For organic deposits, the degree of humification is also recorded on a five point scale as a superscript from 0 (no humification) to 4 (totally humified) (Table 13.7). Hence partly humified but distinct plant structures in a peat comprising 100% turfa would be described as Th^24.

The standard descriptive order suggested for use is: colour, components, humicity, darkness, stratification, dryness, elasticity, boundary contact.

EXAMPLE DESCRIPTION

An example description following the TS scheme is:
- TS Black TH^32, Sh1, Ag1, nig4, stf0, sicc2, elas0, lim sup4.

The above section of the description of peat does not mention the most common peat-related material that is encountered in rocks, namely the coals and seat earths. The description of coals is covered in Chapter 4.11.2 where a simple scheme is suggested, without going into the detail that is required for the non-engineering assessment of coal.

In areas of coal there are two other aspects which are important. The absence of coal might be very important if this is as a result of workings, in which case the description of the materials at and above the anticipated seam level will be critical.

Coals are usually underlain by seat earths, just as most peat deposits are underlain by layers of silt. The thickness of these layers is not usually substantial, but they are of great importance as they are usually of low strength and intensely sheared with polished and striated surfaces and are susceptible to weathering and softening in exposure. They are therefore layers which should be located and described using the standard terms provided in this book.

13.1.3 Transported mixtures of organic soils

Organic materials also occur in transported soils as secondary constituents. These can either occur through erosion of the mire such that fragments of the peat are incorporated into the deposit or, more commonly, from organic accumulations within the fluvial system, in back waters, swamps and wetlands. The source of the organic inclusions within the transported soil is not critical, but recognition of their presence is critical and should be described. Care should be taken to differentiate between discrete inclusions of organic material (pockets of peat) and a disseminated organic content (organic CLAY).

The organic content initially affects only the colour, which is not too important in engineering terms, and causes some odours. As the organic content increases,

Table 13.8 Terms to describe organic soils (BS 5930 withdrawn practice)

Term	Typical colour	Organic content (% by weight)
Slightly organic clay or silt	Grey	2–5
Slightly organic sand	As mineral constituents	1–3
Organic clay or silt	Dark grey	5–10
Organic sand	Dark grey	3–5
Very organic clay or silt	Black	>10
Very organic sand	Black	>5

Table 13.9 Terms to classify organic soils (EN ISO 14688-2)

Term	Organic content (% by dry weight)
Low-organic	2 to 6
Medium-organic	6 to 20
High-organic	>20

EXAMPLE DESCRIPTIONS
- Firm dark grey CLAY with frequent organic inclusions up to 3 mm across
- Black very organic silty fine SAND with rare pieces of partly decomposed wood up to 10 mm

however, it has a marked effect on the behaviour of the soil. Even organic contents of only a few percent (see Table 13.8) can cause raised water contents well in excess of 100% and the plasticity also increases so that such soils plot well below the A line (see Chapter 4.6).

The descriptive terms and the related classification provided in BS 5930: 1999 (pre-Amendment 1) is given in Table 13.8 and provides a useful scheme. This has been removed from Amendment 1 because of conflict with the terms given in EN ISO 14688-2, these terms being indicated in Table 13.9. Which of the classification and terms are used is not too critical. The descriptive terms in either of these tables can be used as appropriate; as the terms are different, the scheme in use will be immediately apparent. For this reason the schemes should not be combined in any way.

13.2 Volcanic soils or rocks

The identification of soil or rock as volcanic will be made by a combination of anticipation from a desk study and from close examination of the nature of the constituent particles in the deposit. A soil that is of volcanic origin can be thus identified by the existence of pumice and scoria. If appropriate, the soil can be washed to separate the grains of volcanic glass, the proportion of which can then be measured. If more accurate identification is needed, a full petrographic examination by a practitioner well versed in these materials will be necessary.

Naming volcanic soils is covered in BS 5930: 1999 in brief terms, but an alternative and more accurate proposal has been put forward by Brown (2007). This is

---

BACKGROUND DEFINITIONS

- Scoria is vesicular glassy lava of basic composition ejected from a vent.
- Pumice is light porous rock formed from gas bubbles, fragile volcanic glasses and minerals and is typically acidic in composition. The term is used to describe both the frothy pyroclastic material and the resultant rock.

---

discussed in Chapter 4.11.4 in the context of rock naming. The description of volcanic soils and rocks is not a great problem, but given the freshness of the material they have a propensity to be unstable and to weather rapidly. This instability, together with the complex and variable processes of formation, means that great attention needs to be paid to the constituents and their arrangement. Description should be carried out in accordance with material specific guidance in the local area in which the investigation is being carried out.

13.3 Loess and brickearth

Loess is a wind blown (aeolian) sediment derived from hot or cold desert areas. It is typically homogeneous, structureless and composed of 50–90% silt particles.

Loess can be remobilised and redeposited by wind, water ('river loess') or solifluction ('head loess') and reworked in situ by cryogenic or pedological processes. In many cases the origin of these deposits cannot be fully explained because of their complex geological history. As there may be a number of depositional episodes within the same sequence, there may be a number of layers with small differences in density, mineral content, structure and particle size.

Brickearth is a term which is often synonymous with loessic deposits and is still used on older geological maps either as 'brickearth', 'head brickearth', 'head silt', 'river brickearth' or 'loam'. However, in the London area these terms have been replaced by a series of 'silt members' including the Langley Silt Member, Enfield Silt Member and Ilford Silt Member. The term 'brickearth' has been used as the material is often suitable for brick manufacture and has been and still is a source of 'brick clay'. Typically, only those deposits with little calcium carbonate content are extracted, which may leave lower beds of higher calcium carbonate content.

Loess is often described as non-stratified, light yellow or brown, sometimes calcareous slightly sandy silt. The particles are mostly silt size (2–63 µm) with lesser and variable amounts of sand and clay. The silt particles may be from glacial and glacifluvial deposits derived from the glaciers grinding up rock to silt-sized particles (rock flour). After drying, these deposits are highly susceptible to wind erosion, which causes winnowing of the silt, transportation and redeposition. Loess grains may be angular or rounded and generally composed predominantly of quartz but also contain feldspar, mica and other minerals. The particles may be sourced locally or from a distance.

Loessic deposit sections may be stable for many years as they tend to contain relatively large pores that drain rapidly so that it is never fully saturated. Vertical fissuring in a prismatic pattern is often seen in the upper part of outcrops; the reasons for this fissuring are unclear but may be due to desiccation (Entwisle, personal communication).

Loess and brickearth deposits are not widespread in the UK, although in other parts of the world there are huge expanses of loess terrain with thicknesses of several hundred metres. It has been estimated that about 10% of the Earth's land surface is covered by loessic deposits. In southern England, these deposits typically occur in discontinuous spreads, about 1–4 m thick, but may be up to 8 m thick in buried erosional channels. It overlies river terrace gravels, chalk, Thanet sand formation and to a lesser extent the London clay formation. Elsewhere in the UK, it overlies Jurassic and carboniferous limestones.

The soils are not particularly difficult to describe as they contain such a narrow range of grain sizes. It is important to describe the variation in grading, in particular the changes in clay content, as small amounts of clay can change the behaviour of the soil from potentially collapsible to stable. Small increases in density can also have the same effect. The presence and form of calcium carbonate content have also been found to be an indicator of collapse and should be described with care. The calcium carbonate occurs as concretions and as discrete, slightly cemented, nodules up to 5 mm in diameter developed in clusters and patches. The logger should, therefore, be concentrating on making sure that the most appropriate samples are recovered, with every effort being made to recover samples suitable for the description of particle size, structure, calcium carbonate content and form and the in situ relative density.

Because of the potential sensitivity of these materials, the description will benefit from being supported by an appropriate and extensive programme of density and laboratory classification testing on suitably sampled cores and blocks.

The environment of deposition means that these soils are laid on an eroded pavement so there can be penetration of the loess into cracks and voids in the pavement surface. This relationship should be described.

14 Made Ground

The final deposit covered in this book is made ground which is the group of soils laid down by man, not by natural processes. This gives rise to levels of non-homogeneity that Nature cannot match. In addition, made ground includes different materials and different associations of materials which, in many instances, are clearly not natural. The skilled logger of made ground therefore needs to be aware of the processes that man has been using to manufacture and to lay the materials, rather than rely on his or her training in the geological sciences.

The initial sections of this chapter (Sections 14.1 to 14.4) provide a framework for the description of made ground in the context of materials, usually soil, deposited by man. There are also a number of man-made materials which are commonly encountered in investigations which are actually part of an engineering structure, such as a road pavement. The guidance that follows in Sections 14.5 to 14.7 is to help those logging structures or structural materials in making a reasonable recognition of the materials and layers and a complete and accurate description.

In order to identify the presence of made ground, the first identification that needs to be made is whether the material is laid by man or naturally deposited. This distinction is probably going to be most difficult in the first of the categories given in Table 14.1. Natural soils that have been re-laid by man may be difficult to identify as such but the logger needs to look for and make use of all the evidence that can be found. This evidence might include one or more of the following:

- results of desk studies which are now carried out routinely at an early stage of investigations and are indeed a requirement of EN 1997-2;
- reading the ground in the context of local ground forms and the shape of the land in terms of unnatural shapes, back filled hollows or excavations;
- location of exploratory hole (e.g. rig on embankment) shows that logging should not be carried out without having visited the investigation position;
- the structure or fabric of the soil which may be lumpy or irregular rather than layered;
- the presence of artefacts (man-made objects) because even only one or two artefacts may be diagnostic. For example, the level of the rare brick fragments or pieces of plastic indicate that all the materials at a higher level could well be made up.

On the other hand, the presence of some of this evidence can be misleading and it is the responsibility of the logger to see through these false trails and to abide by the BS 5930 requirement that the field description is of the ground before the investigation hole was formed. So, BEWARE of the following possibilities of being misled:

- natural alluvial soils can contain artefacts owing to history of human activity in the region. For example, boulders of granite or limestone which were used as ship's ballast were occasionally heaved over the side in port and so are now present in the silts in dock areas, but this does not make those silts made ground;

- similarly, fragments such as brick can be included in alluvial deposits if humans have been around and building for long enough, as is the case in some areas of southern Europe and the Middle East;
- sample contamination added by the drilling crew in forming the hole and taking the samples. It is often difficult to avoid picking up materials from around the rig when the tools are laid on the ground making it possible that samples could include clinker from the rig working area, either from the tools or inadvertently included into the bag sample. The presence of green grass in a sample from 15 m is rather easier to spot and highlights poor practice in operation at that particular rig.

14.1 Types of made ground

The three principal types of made ground are indicated in Table 14.1. The first type is where humans have excavated and replaced the natural soil in a new structure. The nature of the soil will not have been changed but the state may and the structure certainly will have been changed. The second and third categories are materials made by humans, either through processing natural materials or through actually manufacturing the material. Where these materials have a chemical and/or physical similarity to natural soils they can probably be sampled, tested and described

Table 14.1 Types of made ground

Category	Nature of made ground	Examples
Natural soils excavated and re-deposited by humans; see example in Figure 14.1	• Use normal approach and terminology as outlined in this book. • Usually not too much of a problem • Can be sampled and tested in accordance with normal geotechnical test procedures.	• Embankment fill • Colliery spoil (coarse discard) • Drainage layer, e.g. gravel
Man-made materials describable and testable geotechnically: see example in Figure 14.2	• Can frequently be described using normal approach and terminology as above. • Can be sampled and tested in accordance with normal geotechnical test procedures.	• Crushed rock backfill • Washed materials such as colliery spoil (fine discard), mine tailings from non-coal mines • Crushed concrete • Pulverised fuel ash (PFA)
Man-made materials NOT readily describable and testable geotechnically; see example in Figure 14.3	• Materials defy description in any standard manner. • Include a range of exotic materials and artefacts. • Often not testable in the field or in the laboratory. (cannot measure strength of a bicycle frame or liquid limit of plastic bag). • The description is ALL YOU HAVE.	• Landfill • Demolition rubble (including frames, slates etc) • Fly tipped materials

Figure 14.1 Embankment fill showing clods of different coloured clays (and a throughgoing shear surface).

Figure 14.2 General fill comprising processed gravels as well as bricks and burnt shales.

using the approach for natural soils described so far in this book. Once the material is physically or chemically different from natural soils, this system will break down and alternative descriptive approaches need to be found. In the guidance below, some pointers are given as to how each of these categories of material can be described.

Figure 14.3 Landfill showing heterogeneity and large pieces of materials.

There is often a distinction in engineering practice between:

- fill which was placed under engineering control, and
- made ground which was placed without any noticeable control.

These engineering definitions are not in agreement with those used by the British Geological Survey on published maps. The definition used there is:

- Fill was placed below original ground level (recently this has been termed infilled ground).
- Made ground was placed above original ground level (more recent maps will show this as made up ground).

There is also a range of less prosaic man-made materials which are generally related to a single manufacturing process, possibly only at one or two plants. Others relate to a broader geographic area but remain regional owing to the geological conditions. Some examples of such terms are given below:

- Blaes arises from combustion of oil shale to extract paraffin in Scotland. It is typically an orange shale gravel which is commonly used as road sub-base.
- Galligoo is a by product from the manufacture of soap and soda particularly in the Widnes/Runcorn area. The calcium sulphate ('black gunge') can combine with saline water (from the local salt rock deposits) to create a toxic leachate which gives off quantities of hydrogen sulphide gas.
- Burgy is a waste that arises specifically from float polishing glass at the Pilkington plants in St Helens and Doncaster. The material comprises a mixture of sand and jeweller's rouge, deposited in bunded lagoons up to about 15 m deep. In some areas the material is seen as a highly alkaline brown liquid emanating from the base of 'slag heaps'.
- Chalk whitings include a range of materials derived from chalk which has been ground, partly baked or fired. The products are used in a wide range of

applications including whitewash, putty (mixed with linseed oil) and as whitener in ceramics. Only found in areas of chalk workings.

There are surely many other examples of such manufacturing process-specific by-products which can be encountered as made ground. These materials can have unusual properties owing to their grain size or chemistry.

14.2 Identification of made ground

The aim of describing made ground is to identify those constituents which are stable and will behave conventionally, from those that are problematical, whether this is due to instability, obstruction or hazard. One suitable approach is that included in HA Test 710 although this is aimed specifically at assessment of recycled aggregates.

Following this test method, a suggested two stage procedure for the description of made ground is as follows. The first stage is to describe the overall behaviour of the made ground and the second is to identify the materials that make up the 'deposit'. Not all made ground will justify both approaches, although all made ground descriptions should in principle use one of these approaches.

The first approach is to divide the material into the size fractions as outlined in Chapter 4.3. The three divisions of the very coarse, the gravel and the finer fraction (comprising the sand and fine soil sizes) should be described in terms of how they will behave and combined to indicate the proportions of each. In other words the overall behaviour and grading of the made ground should be assessed following the guidance given throughout this book. This description will need to incorporate evidence or observations about how the material behaves on the bench, in the spoil heap or in the sides of the excavation.

TIP ON CLASSIFYING THE COMPONENTS OF MADE GROUND

- A further refinement in considering the variety of materials within made ground is to measure the proportions of the components. The basis for this proposal is the Highways Agency Clause 710 Test for recycled aggregates (Highways Agency, 2007a).
- In this test, samples of materials are analysed to determine their composition. The test comprises hand sorting after all particles finer than 8 mm and coarser than 63 mm are removed. This leaves the medium and coarse gravel size fractions.
- For the purposes of made ground description the range of grain sizes that are included would need to be extended to include the fine gravel and coarse sand as well as the cobbles.
- In the HA 710 test the particles are then sorted by hand into six categories
  - asphalt (Class A)
  - masonry (Class B)
  - concrete (Class C)
  - lightweight blockwork (Class L)
  - unbound aggregate (Class U)
  - other materials, glass, metal, wood, plastics, etc. (Class X).
- On completion each category is weighed and the weight of each fraction reported as a percentage. The tests can be repeated in duplicate.
- For the test on recycled aggregate it is required that each category fraction must have no less than 500 particles and this will necessitate about a sample with about 100 kg of 63 mm down.

The different constituents of the made ground should then be assessed. One or two bucketfuls of excavated material should be placed on the ground away from the trial pit and spread out with a shovel. The area covered by the sample should then be divided into a grid of any sensible size (a spacing of between 200 mm and 400 mm is suggested). The proportions of each type of material (see below) within each square should be assessed by volume. The results from several squares would be averaged and the results reported as volumetric percentages.

The groups of material that should be assessed could include:

- natural soil materials;
- man-made materials subdivided into groups such as:
  - asphalt
  - concrete
  - masonry, blockwork and brickwork
  - glass, metal, plastics
  - wood, paper
  - burnt products such as slag, clinker and ash including evidence by strong colours;
- large objects and obstructions which may be critical to piling operations;
- hollow objects, compressible/collapsible objects or voids which could be important to ground stability;
- chemical wastes and dangerous or hazardous materials;
- decomposable materials with note on degree of decomposition; a simple scale of decomposition can be used, noting whether the material is not decomposed (materials fresh), moderately decomposed (materials stained) or highly decomposed (materials falling apart).

In addition, information should be included on:

- any dating possible on papers or other artefacts
- signs of current or recent heat or combustion
- smells or odours (see below)
- layers and their inclination to inform on mode of tipping, whether ponded, end tipped, spread or stockpiled
- origin of materials if identifiable.

Even if the procedure suggested above is not followed in full, the outcome of any description should provide a volumetric estimate of the made ground constituents. Some examples of descriptions that can arise in made ground are given below. These examples illustrate the flexibility in word order which is both necessary and allowed to accommodate the wide variety of materials concerned. The man-made nature can be emphasised by highlighting the artefacts in the description. Perhaps:

- COBBLES of slag, GRAVEL of brick, tiles…
- MADE GROUND composed of…

The flexibility in approach in these examples is completely intentional so that the logger can achieve the aim of good communication with clarity and no ambiguity. The logger should also try to avoid excessive detail. Whilst a description always needs to be complete and accurate, loggers should ask themselves whether they

EXAMPLE DESCRIPTIONS

- MADE GROUND comprising plastic bags, window frames, garden refuse, newspapers (1964)
- MADE GROUND: dense brown sandy GRAVEL with occasional tiles, wire, glass, tyres
- Soft grey sandy CLAY. Rare gravel size brick fragments (MADE GROUND)
- Firm yellow brown slightly sandy CLAY with clods (up to 200 mm) of firm to stiff orange CLAY (EMBANKMENT FILL)
- Dense white sandy limestone GRAVEL with medium generally subangular cobble content. (SUB BASE)
- MADE GROUND comprising:
  - 50% pockets up to 1.0 by 0.4 m of black partially decomposed paper, newspapers (1962), garden refuse and ash
  - 25% multicoloured (bright colours) clays (possible dyes)
  - 20% concrete slabs up to 1.5 by 0.2 m lying at 45°
  - 5% 1 No 200 litre drum slightly corroded, apparently empty, no labels, hydrocarbon odour.

believe the made ground is likely to be similar a metre or two away, or not. If it is not, too much detail in the description may be counterproductive.

14.3 Odours

The description of odours is an area where little guidance has been provided at a level appropriate to soil description. The main difficulty is that the description of odours is highly subjective. In assessment of odours in manufacturing or the environment, it is usual to set up odour panels of up to about ten people who rank the smell in terms of intensity, quality and acceptability. This is clearly not a practical approach in day-to-day soil description.

There is a further difficulty in assessing odour in that the nose is very good at accommodating compounds that it thinks might be harmful; once the olfactory sensor cells have been stimulated by a particular smell, they stop signalling to the brain.

It is generally accepted that there are seven types of smell corresponding to the seven types of smell receptors in the olfactory-cell hairs and these are given in column 1 of Table 14.2. For the purposes of description of odours in made ground, it seems that an eighth is needed to accommodate the most common odours

Table 14.2 Terms for description of odours

Category	Descriptive terms
Camphor	Bitter; mothballs; acrid
Musk	Penetrating; pungent
Floral	Wide range of terms, not likely to be used often in made ground
Peppermint	Sweet; minty
Ether	Solvent; acetone; medicinal
Vinegar	Sharp; acetic; pungent; rancid
Putrid	Rotten egg; rotten cabbage; fishy; disagreeable, sweet; sulphurous
Hydrocarbon	Petrol; diesel; oil; asphalt; tar

encountered in made ground, namely the hydrocarbons. The terms that can actually be used for odour description are given in column 2 of Table 14.2; these are suggestions that are not particularly linked to column 1 but allow more precision than the terms in column 1 on their own.

The description of odours is not a significant problem area in description. The logger should report any smells noticed using the terms above for guidance. If any particularly strong smells are apparent, in the sample or in a container, care should be taken to avoid smelling this too closely or deeply.

14.4 Definitions of some combustion products

The range of made materials which are the result of a variety of combustion processes is given below. Often these materials are physically unstable and usually contain concentrations of metals and polynuclear aromatic hydrocarbons. The definitions below have been found to be workable compromises:

- ASH is sand or silt size by definition, so descriptions do not need to use two terms such as 'ash sand'. It is not possible by this definition to have 'gravel size ash' although cinders can be gravel size but readily crush down. The ash material can include sand size particles of unburnt coal.
- CLINKER is gravel size or larger by definition so descriptions do not need to, but can, use 'clinker gravel'. It is not possible by this definition to have 'sand size clinker'.
- SLAG comprises materials which are fused or poured as liquid or scum or froth, of any size or shape and will be at least strong. If these materials occur in blocks or layers, they can present substantial difficulties for penetration by drilling or excavation. Slag can be pelletised, expanded or crushed for reuse in construction as aggregate.

14.5 Description of concrete or macadam

Concrete and macadam are the terms used in description of the manufactured materials that investigations most commonly encounter. In European and American practice, these materials are called cement concrete and asphalt concrete, respectively. In site investigation practice, these terms can be used, or more simply using the terms concrete or macadam, respectively, should be acceptable. Some background on the naming of macadam materials is provided below. Given the similarity of these materials, even to the concrete association in the name, their description can be considered together here. Examples of concrete cores are shown in Figures 14.4 and 14.5.

The description of concrete and macadam is in many ways similar to the description of a rock. The material is usually matrix supported, the fine grained matrix containing gravel size clasts (coarse aggregate). The matrix is either the cement paste or the bituminous material, which in both cases can include a sand content. It is therefore logical to follow the same principles and even word order in describing these materials, as follows. The use of additional sentences to describe the details of the concrete is usually necessary.

- Strength (using the rock strength scale)
- Colour
- CONCRETE (or MACADAM)

Figure 14.4 Core of cement concrete showing reinforcement steel to either side of the coin.

Figure 14.5 Cores of asphalt concrete of a flexible road pavement showing multiple construction layers (photograph courtesy of Soil Mechanics).

- Aggregate
- Voids
- Deleterious substances if present
- Reinforcement if present (as shown above)
- Other information, e.g. stained fractures
- Formation name, e.g. FOOTING, WING WALL, TRENCH BACKFILL, WEARING COURSE.

The use of formation names such as CONCRETE or REINFORCED CONCRETE is not helpful as that much information is already included within the description. The formation name used should add something to the description.

The fine aggregate (sand size) is not separately described as it cannot easily be distinguished from the cement paste and in field description visual distinction

BACKGROUND HISTORY ON NAMING OF BLACK TOP MATERIALS

The form of road surfacing originally proposed by McAdam was to use the native soil (sub-grade) covered by a road crust to protect the soil underneath from water and wear. He used 2-inch broken stones in a layer 6–10 inches deep to support the road and the traffic upon it and depended on the road traffic to pack it into a dense mass, although for quicker compaction, a cast-iron roller could be used.

This basic method of construction, sometimes known as 'water-bound macadam', required a great deal of manual labour but resulted in a strong and free-draining pavement. Roads constructed in this manner were described as 'macadamised'.

With the advent of motor vehicles, dust became a serious problem on macadam roads. The vacuum under fast-moving vehicles sucks dust from the road surface, creating dust clouds and a gradual pulling apart of the road material. This problem was rectified by spraying tar on the surface to create 'tar-bound macadam' (tarmac).

Hooley's 1901 patent for tarmac involved mechanically mixing tar and aggregate prior to placement and then compacting the mixture with a roller. The tar was modified with the addition of small amounts of Portland cement, resin and pitch.

The Macadam construction process became obsolete owing to its high manual labour requirement. However, the tar and chip method, also known as bituminous surface treatment (BST), remains popular today.

Early tar sprays used coal tar derivatives, but as petroleum production increased, the by-product asphalt became available in huge quantities and supplanted coal tar because it had reduced temperature sensitivity. Such materials are generally called asphalt or bitumen surfacing. There is an important chemical difference between the two sources of tar. Coal tars are very high in poly aromatic hydrocarbons (PAH) (>5000 mg kg^{-1}) whilst asphalt is less contaminated, with PAH levels of about 50–100 mg kg^{-1}.

Modern practice in naming these materials is as follows:

- Tarmac is strictly those materials made from coal tar and its derivatives.
- Bitumen or asphalt is made from oil and is termed asphaltic concrete (as distinct from cementitious concrete).
- The distinction between tarmac and asphalt requires specialist knowledge and should not be attempted in the normal course of investigations.

of the sand or any fillers or resinous materials added to the matrix will not be possible.

The example descriptions given above represent the normal level of detail that will be sufficient for most projects. There will also be projects where merely identifying that concrete or macadam is present will suffice and no detailed description is

EXAMPLE DESCRIPTIONS

- Medium strong light grey CONCRETE. 40–50% aggregate of angular to rounded fine to coarse basalt gravel with rare cobbles. 10% small voids. Rare decomposed plant remains (<10 by 3 mm). Concrete friable for 20 mm around plant remains (EXCAVATION BACKFILL).
- Strong light grey CONCRETE. 70–80% aggregate of angular to subangular fine to coarse basalt and chert gravel with rare cobbles. 2% small voids. 5 mm and 10 mm reinforcement bars of mesh at 100 mm spacing. (RETAINING WALL).
- Weak black MACADAM. 70–80% aggregate of angular fine to medium limestone gravel. 5% small and medium voids. (BINDER COURSE).

required. Some projects, however, may require additional detail, usually if a problem or failure is being investigated, or if re-use or development of the concrete or macadam is proposed. The additional details that may be required are given below. Description of this greater level of detail should only be carried out when it is specifically needed for the project and an instruction has been given by the client through the project manager.

14.5.1 Aggregate content

Detailed description of surface texture and particle shape of the aggregate can be carried out in accordance with BS 812: Part 105: 1989 using the terms shown in Tables 14.3 and 14.4. The shape descriptors given in Table 14.4 can be used together with the terms provided in Figure 4.4.

Figure 14.6 Example of logging concrete in the field requiring examination of different types of concrete and reinforcements.

Table 14.3 Terms for description of the surface texture of aggregate

Surface texture	Characteristics	Examples
Glassy	Conchoidal fracture	Flint, vitreous slag
Smooth	Water-worn or smooth owing to fracture of laminated or fine-grained rock	Gravels, chert, slate, marble, some rhyolites
Granular	Fracture showing more or less uniform rounded grains	Sandstone, oolite
Rough	Rough fracture of fine or medium grained rock containing no easily visible crystalline constituents	Basalt, felsite, porphyry, limestone
Crystalline	Containing easily visible crystalline constituents	Granite, gabbro, gneiss
Honeycombed	Visible pores and cavities	Brick, pumice, foamed slag, clinker, expanded clay

Table 14.4 Terms for description of the particle shape of aggregate

Particle shape	Description	Examples
Rounded	Fully water-worn or completely shaped by attrition	River or seashore gravel, desert, seashore and windblown sand
Irregular	Naturally irregular or partly shaped by attrition and having rounded edges	Other gravels, land or dug flint
Angular	Possessing well-defined edges formed at the intersection of roughly planar faces	Crushed rocks of all types, talus, crushed slag
Flaky	Material in which the thickness is small relative to the other two dimensions	Laminated rock
Elongated	Material, usually angular, in which the length is considerably larger than the other two dimensions	
Flaky and elongated	Material having the length considerably larger than the width and the width considerably larger than the thickness	

Table 14.5 Terms for description of voids

Term	Definition of size in any direction
Small void	0.5–3 mm
Medium void	3–6 mm
Large void	Larger than 6 mm
Honeycombing	Interconnected voids arising from, for example, inadequate compaction or lack of cement paste

14.5.2 Voids

An estimation of void content is made based on the percentage by volume. The size of voids is described using the terms in Table 14.5 (after BS 1881: Part 120: 1983).

14.5.3 Deleterious substances

Any materials which look as though they should not be there should be noted. These materials can have a deleterious effect on the performance of the concrete or macadam:

- impurities which interfere with the process of cement hydration, e.g. organics;
- coatings on coarse aggregate of clay or silt which prevent a good bond developing between aggregate and cement paste such that individual particles can be plucked out of the matrix;
- individual particles which are unsound or weak, e.g. coal, mudstone, clay pockets, wood, shells, iron pyrite and gypsum.

Products of alkali aggregate reaction (AAR), alkali silicate reaction (ASR) or thaumasite sulphate attack (TSA) can be seen as rims of soft material such as paste or gel around aggregate, loosening the bond to aggregate particles.

14.5.4 Reinforcement

The details of any reinforcement present should be described. Information that can be recorded includes:

- bar/mesh strand diameter and spacing (if visible)
- surface texture (e.g. corrosion, pitting)
- whether ribbed or unribbed.

If no reinforcement is observed in cement concrete, a positive statement to this effect in the description will often be helpful.

14.6 Pavement material types

Definitions of the macadam or concrete layers that go into the construction of a road pavement are given in Table 14.6 for background information. This information should not influence the description of the materials seen, but field loggers should find this summary a useful aid in identifying those aspects of the materials that help distinguish between the different layers that are visible.

Bituminous road pavements are classified in terms of their designated purpose/position in the construction as shown in Table 14.7. This profile is for flexible pavements (asphaltic concrete surface). In rigid pavement construction (cement concrete surface) there may be a surface layer of high density concrete and base layers of concrete which may be reinforced or unreinforced and which rest directly on the sub base.

It should be noted that whilst binder and base courses have distinct uses, these and other layers may be visually indistinguishable and the terms given in Table 14.7 should only be used when it is clear which layers are present. This will most often be

Table 14.6 Bound material types

Hot rolled asphalt (HRA)	Hot rolled asphalt is a gap graded material with coarse aggregate held in a binder-rich matrix of sand and filler. Occasionally quarry fines are used instead of sand which makes them difficult to distinguish from macadam. HRA is recognisable as having well distributed coarse aggregate embedded in a fine textured dark matrix. The aggregate is usually hardstone for surface courses. Surface course HRA usually has 20 mm, sometimes 14 mm, chippings embedded in the top surface. HRA is also used in binder and base courses.
Macadam (M)	Macadams are distinct from HRA in having continuous (wide) gradings and crushed rock fines. They are sometimes made with sand fines and are then difficult to distinguish from HRA. Macadams are usually less dark than HRA as they are less binder rich. Macadams are used in surface, binder and base courses. Surface courses are usually made with hardstone. Binder courses and bases usually have larger nominal sizes (20 mm, 28 mm and 40 mm) than surface courses. Limestone surface course is often used in thin layers as a regulating course to smooth out old surfaces prior to adding a new surface course.
Dense bitumen macadam (DBM)	Generic name used to describe most macadams.
Porous asphalt (PA)	A very low fines aggregate and fairly high binder mix, designed to be porous and allow surface water to drain through it, more voids than SMA (see Table 14.7).

Table 14.7 Material descriptions of bound road layers

Type and abbreviation	Description
Surface course (SC)	The top running surface of a highway identifiable as having hardstone aggregates and a coarse aggregate usually 14–20 mm. Surface courses may be buried underneath successive resurfacings.
Stone mastic asphalt (SMA) (TSCS)	A surface treatment consisting of a gap graded aggregate with a high binder content. Recognisable by the low fines content and binder-rich nature. Can also be used to cover a number of proprietary surface treatments; these are referred to as thin surface course systems (TSCS). Usually has an open, negative textured, surface.
Base and binder courses (B or BC)	A layer to spread the load imposed on the surface course to the base. Usually laid more accurately than base to produce a 'smooth' surface course. The aggregates used for binder courses are usually limestones but may differ depending on local supplies. Nominal aggregate size is usually 20 mm but may be 14–28 mm. Binder and base have distinct uses but may be visually indistinguishable and the terms should only be used when describing a modern construction where the conventional course structure has been used.
Base (B)	The main load-bearing element within the bound road pavement structure and as such is usually the thickest and the stiffest layer. Dense bitumen macadam is the most common, although asphalt is also used. In some designs the base is replaced with a cement-bound material (previously referred to as lean mix or lean concrete).
Regulating course (RC)	Used to smooth out old or planed surfaces prior to laying a new surface course. Limestone surface courses are often used in thin layers, often dense bitumen macadam with a nominal aggregate size of 3–20 mm.
Sub base	The lowermost layer of the pavement construction which spreads the load of the bound layers above and acts as a drainage layer. All conventionally designed roads have a sub-base layer normally comprising a well graded coarse soil with a minimum of 10% sand and no cobbles (typically 'sandy GRAVEL') and may contain cement or lime stabiliser. Type 1 sub base is generally crushed 'sandy GRAVEL' (up to 30% sand) whereas type 2 may be crushed or natural 'sandy GRAVEL' or 'SAND and GRAVEL'.
Capping	An initial cheaper stabilising layer of construction on a weak subgrade and so may often be found above a separator or reinforcing fabric. Similar to sub base but may contain fine soil constituents, thus 'clayey sandy GRAVEL' and may include cement or lime.
Subgrade	The existing natural or man-made soil on which the pavement was constructed, possibly altered by lime stabilisation which is unlikely to be visible except by a top drier/stiffer layer. The formation name given would normally reflect the underlying sequence.
Definitions	• Coarse aggregate – rock material of gravel size • Fine aggregate – crushed or naturally occurring sand size particles • Crushed rock – quarried rock materials crushed to produce an aggregate with a specified grading • Binder – usually penetration grade bitumen produced from crude oil • Filler – material of silt size or finer, usually quarry dust • Hardstone – aggregate generally used for surface course materials, can be any crushed rock other than limestone and natural gravels (as gravel above).

when describing a modern standard road construction, although this is less likely to occur within a site investigation.

It is important to remember that as a highway undergoes maintenance, new material may be placed on top of the previous material. As such it is possible to find several surface courses in any one core; the material designation does not change as it is overlaid.

14.7 Description of brickwork

The word order for describing brickwork is also essentially the same as for concrete, but there are variations owing to the 'placed' rather than the 'poured' method of construction. The bricks themselves and the mortar can be clearly seen in cores (see Figure 14.7) and both require description.

The word order is as follows:

- Strength (using the rock strength scale)
- Colour
- BRICKS
- Size (quoted in brackets if known)
- Strength (using the rock strength scale)
- Colour
- MORTAR
- Composition (where appropriate)
- Bonding
- Thickness
- Other information (e.g. voids, deleterious substances, etc.)
- Formation name, e.g. STRIP FOOTING, PARAPET WALL.

The bonding of the mortar to the bricks can only be described in qualitative terms. A suggested terminology is given below. The quality and diameter of coring can affect the bonding:

- Well bonded mortar remains bonded to brickwork on retrieval.
- Poorly bonded mortar is not bonded to brickwork after retrieval.

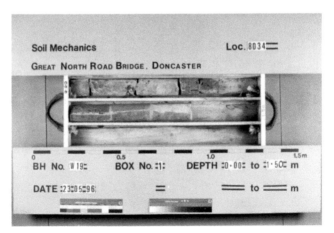

Figure 14.7 Horizontal cores of brick work (exposure 1.0 m long) (photograph courtesy of Soil Mechanics).

BACKGROUND ON BRICK SIZES

There are two common sizes of bricks. Special types of bricks such as engineering or refractory bricks often have different dimensions:

Old stock (Imperial) bricks: 8.5″ × 4″ × 2.5″or 216 × 102 × 64 mm
New stock (metric) bricks: 210 × 100 × 60 mm

EXAMPLE DESCRIPTION

Moderately strong yellow BRICKS (216 × 102 × 64 mm). Strong grey MORTAR well bonded, 10 mm thick (TUNNEL LINING).

14.8 Logging cores of manufactured materials

Where coring is carried out in concrete, macadam or brickwork, whether this is an integral part of the investigation or merely creating access though the slab or pavement for investigation of the materials below the pavement or slab, the cores should normally be described as outlined below. There will be projects where no detailed description is considered appropriate.

The procedure for logging of cores of concrete, macadam or brickwork is essentially the same as for rock cores. However, as these are manufactured materials, there is more emphasis on the materials present than the mass aspects such as the fractures. Logging should be carried out as follows:

- Place the core on a suitable surface.
- Measure the depths of the visible changes in the construction courses.
- For bound road material layers, describe the materials within each course, which have been identified in accordance with the guidance in Tables 14.6 and 14.7.
- Record the thickness of each course.
- Record the rock type of the coarse aggregate of each course in terms of limestone, flint or chert, or other lithology.
- Record the start and finish depths of any cracks, their primary direction and maximum and minimum widths, where possible.
- Record the size range and position of any voids.
- Record the presence of any soft or binder rich material.
- Record any smell of tar or kerosene.
- Record any evidence of loose aggregate.
- Record the date logged, the name of the logger, the core identity and the job number.
- Record any other information required by the client.

In terms of recording fractures, obviously any cracks, boundaries or defects should be noted, but fracturing in the sense of jointing or fissuring is not sensibly relevant. Therefore, in concrete, macadam or brickwork cores:

- Total core recovery (TCR) should be recorded.
- Solid core recovery (SCR) can be recorded where useful, but this is not compulsory.

- Rock quality designation (RQD) and fracture spacing (If) should not be recorded.

Cores should be photographed following logging (unlike rock cores) as shown in Figure 14.5 and Figure 14.7. Cores should be arranged so that any damage or cracking is visible. Adjacent core pieces should be rotated to match the piece above and any layers of loose stripped material should be arranged to match the appropriate depth interval.

15 Classification Schemes

The previous chapters have covered the description of soils and rocks. The classification of soils and rocks is the next step in the process of investigating the ground and this incorporates the process of assembling a ground model, which provides a framework and shorthand for communication of the findings of the logging into the design process. There are a number of levels at which this classification can be carried out. In a sense the division of the field descriptions into strata units is the first step in this process, in that the boundaries between units are placed at the point at which the field logger considers that the material properties have changed and this boundary is likely to represent a change to another unit within the classification.

As identified in the EN ISO standards, the description and classification of soils are fundamentally different activities.

- EN ISO 14688-1 sets out the precept in that the 'general identification and description of soils is based on a flexible system for immediate (field) use by experienced persons, covering both material and mass characteristics by visual and manual description'. This is therefore a procedure carried out in the field without the use of test results.
- EN ISO 14688-2 sets out classification principles to 'permit soils to be grouped into classes of similar composition and geotechnical properties, and with respect to their suitability for geotechnical engineering purposes, such as:
  - foundations
  - ground improvement
  - roads
  - embankments
  - dams and
  - drainage systems'.

The fundamental requirement for the use of the classification principles incorporated in EN ISO 14688-2 is the availability of test results.

In many countries it is considered that the first step in the process of recording the soil and rock types is the classification carried out by the lead driller in the field. The view that the driller's description is a classification, may arise from translation, but this first step in the process of description and classification is laid out in Table 15.1, which is derived from Annex A of EN ISO 14688-2. The similarity of this table to that presented as Table 4.3 and discussed throughout Chapter 4 is plain to see and confirms the overlap between description and classification within the investigation process. It is only when concentrating on soil description, as in this book, that a clear distinction between description and classification, as laid out in the introductions to EN ISO 14688-1 and EN ISO 14688-2, needs to be made.

Table 15.1 Initial categories for the description and classification of soils

Criterion	Soil type	Definition	Distinction of soil groups based on manual properties		Further subdivision as appropriate
Wet soil does not stick together	Very coarse	Most particles >200 mm	Boulders	Boulder mixtures with finer material	Classification should be in accordance with project requirements.
				Cobble/boulder mixtures	
		Most particles > 63 mm	Cobbles	Cobble mixtures with finer material	
	Coarse	Most particles >2 mm	Gravel	Gravel mixtures with sand and fine soil	Particle size grading Particle shape Lithology Particle strength Relative density (Permeability)
		Gravel and sand can be subdivided into fine, medium or coarse		Sand/gravel mixtures	
		Most particles >0.063 mm	Sand	Sand mixtures with gravel and fine soil	
Wet soil sticks together and remoulds	Fine	Low plasticity dilatant	Silt	Silt mixtures with coarse soil	Plasticity Water content Strength, sensitivity Compressibility, stiffness (Clay mineralogy)
		Differentiation between silt and clay on basis of hand tests		Silt/clay mixtures	
		Plastic non-dilatant	Clay	Clay mixtures with coarse soil	
Dark colour, low density	Organic	Accumulated in situ	Peat	Peat can include inorganic soil.	Classification should be in accordance with project requirements.
		Transported mixture	Inorganic soils	Organic/inorganic soil mixtures	
Not naturally deposited	Made ground		Made ground	Relaid natural materials	As for natural soils
				Man-made material	As for natural soils
				Man-made material	Classification should be in accordance with project requirements.

It is therefore logical to consider that classification of the ground at a site should be considered to proceed progressively through the various stages of an investigation. The initial stage would comprise identification of the geological succession and the units and subunits within that succession using the approach outlined in Table 15.1. This identification of the formations and their character is part of the description and so would normally be expected to appear on field logs (see Chapter 9).

The second stage might be to follow common practice in many investigations and construct a site-specific classification appropriate to the geology and the engineering problem. This will be based on the geological sequence, but will also be enhanced by those engineering aspects of the sequence that are relevant to the particular nature and demands of the project. Such a classification could be along the following lines and can be made as complex or as simple as desired:

- a. Alluvial deposits:
  - a1 soft clay
  - a2 peat
  - a3 basal sands
- b. Glacial deposits:
  - b1 laminated clay (lacustrine deposits)
  - b2 stiff clay (glacial till)
  - b3 sand and gravel (fluvio-glacial deposits)
- c. Bedrock:
  - c1 weathered bedrock
  - c2 intact bedrock (Carboniferous deposits)
  - c3 intrusive igneous rocks (Tertiary).

A classification such as this is useful right from the earliest stages of any investigation as it provides a framework for understanding the ground and for the continual refinement of the ground model by allowing a straightforward response to the findings of individual investigation activities. The classification is also of great use in the scheduling of field and laboratory tests. Naturally, the use of such a site classification also simplifies presentation and discussion of the ground and the engineering aspects to others in the design team. The placing of soil or rock strata into a limited number of groups, possibly with shorthand identifiers, on the basis of the geology, the description and classification, using early test results such as grading and plasticity is a key stage in the construction and evolution of the ground model.

The soil or rock classification therefore starts to appear within the ground investigation report text, plans and cross sections, but this is not necessarily reflected back onto the field logs. Where the classification usefully identifies, say, the geological succession, this information should be included on the logs. Some of this classification information might often be a late addition, rather than something identified in the field, as it may be based on test results, on specialist input, on the geological conditions or on the nature of the engineering problems being addressed by the investigation.

Classification is therefore a process that is undertaken after the fieldwork and when tests have been carried out and the results are available. There is obviously a grey boundary between these stages as some tests can be carried out in the field, or at least during the fieldwork programme. However, there remains a clear distinction. Description is carried out in the field and reported on field logs, whereas classification is carried out at some later stage and the results do not appear on the field logs.

The classifications referred to above relate to soils and rocks in combination, on a site-specific basis. However, there is no common approach to national or regional classifications based on both soils and rocks that is widely available or in use. The following sections therefore deal with the classification of soils. The classification of rocks is covered on its own in Section 15.4.

There are many classification systems available for use, many of these having been derived for national use to suit the nature of the soils and rocks in the region or for the particular engineering problems in hand. No attempt will be made to cover all of these classification systems here. Instead, an introduction to the principles that lie behind the most commonly used type of classification is given below. However, one of the most useful aspects of widely used classification schemes is the extrapolation of prior knowledge on ground behaviour onto new sites.

15.1 Classification according to EN ISO

A number of classification options are given in EN ISO 14688-2 based on individual measurements made in laboratory tests. These overlap with some of the terms used in soil description as summarised in Table 15.2. This is an area of potential confusion. The classification terms are based on results from tests such as plasticity, density index and consistency index. As a general rule, these classification terms are for use in the description of the site and the underlying ground and would not be expected to appear on field logs arising from the exploratory holes.

The recommendations in the EN ISO in the matter of classification are not particularly clear. However, the document does set out a useful and extensive list of parameters that can be used in classification. In practice the most common classification parameters used are the soil grading, density or strength, plasticity, water content and soil chemistry.

15.2 International classification systems

The classification system that is most widely used internationally is the Unified Soil Classification System (USCS) which describes the texture and grain size of a soil, that is, the nature of the soil. The classification system can be applied to most unconsolidated materials and is given in Table 15.3.

Extensions of this classification into general engineering characteristics such as suitability for embankments dams, in compaction and for foundations are widely used in many countries. Table 15.4 provides typical examples of such extensions of the system (compiled from a variety of sources including AASHTO, NAVFAC and DIN).

It is apparent that the USCS and this type of extension classification system only covers grading and plasticity, that is, the nature of the soil. No account is taken of the state or water content of the soil and thus the strength and stiffness.

A less wide ranging type of classification is given in Annex B (Informative) of EN ISO 14688-2 which is based on grading alone. The triangular diagram has as its corners Gravel, Sand, and Silt and Clay together. This diagram is represented in Figure 15.1. This particular approach suits the soils in those countries that have adopted this approach such as in Scandinavia. Although this is also a useful

Table 15.2 Classifications in EN ISO 14688-2

Parameter	Terms	Definitions and comment	Relevant section
		Nature of soil	
Particle size fractions	Terms for low, medium and high boulder and cobble content	Classification of very coarse soils only is provided. Classification of finer soil is based on plasticity and grading.	4.3, 9.1
Particle size grading based on shape of grading curve	Multi graded Medium graded Even graded Gap graded	$c_u > 15$, $1 < c_c < 3$ $6 < c_u < 15$, $c_c < 1$ $c_u < 6$, $c_c < 1$ c_u high, c_c any value (c_u = uniformity coefficient, c_c = coefficient of curvature)	
Plasticity	Terms of low, intermediate and high plasticity without definition	Terms are based on plastic limit so there is conflict with those based on plasticity chart and thus liquid limit (see below).	4.7
Organic content	Low organic Medium organic High organic	Organic content 2–6% Organic content 6–20% Organic content >20%	13.1
		State of soil	
Density index	Very loose Loose Medium dense Dense Very dense	Density index <15% Density index 15–35% Density index 35–65% Density index 65–85% Density index <85% No terms provided for classification based on SPT N values.	
Strength of fine soils	Extremely low Very low Low Medium High Very high Extremely high	c_u <10 kPa c_u = 10–20 kPa c_u = 20–40 kPa c_u = 40–75 kPa c_u = 75–150 kPa c_u = 150–300 kPa c_u >300 kPa	5.3
Consistency index	Very soft Soft Firm Stiff Very stiff	Consistency index <0.25 Consistency index = 0.25–0.50 Consistency index = 0.50–0.75 Consistency index = 0.75–1.00 Consistency index >1.00	5.2
Other suitable parameters may be used	Dry density Clay activity Mineralogy Saturation index Permeability Compressibility Swelling Carbonate content	No definitions or terms are provided for these additional classification options.	

Table 15.3 The Unified Soil Classification System

Major divisions			Group symbol	Group name
Coarse grained soils; more than 50% retained on 0.075 mm sieve	*Gravel* >50% of coarse fraction retained on 4.75 mm sieve	Clean gravel <5% smaller than 0.075 mm sieve	GW	Well graded gravel, fine to coarse gravel
			GP	Poorly graded gravel
		Gravel with >12% fines	GM	Silty gravel
			GC	Clayey gravel
	Sand ≥50% of coarse fraction passes 4.75 mm sieve	Clean sand	SW	Well graded sand, fine to coarse sand
			SP	Poorly-graded sand
		Sand with >12% fines	SM	Silty sand
			SC	Clayey sand
Fine grained soils; more than 50% passes 0.075 mm sieve	Silt and clay liquid limit <50	Inorganic	ML	Silt
			CL	Clay
		Organic	OL	Organic silt, organic clay
	Silt and clay liquid limit ≥50	Inorganic	MH	Silt with high plasticity, elastic silt
			CH	Clay with high plasticity, fat clay
		Organic	OH	Organic clay, organic silt
Organic soils			Pt	Peat

Key to abbreviations

G	gravel	P	poorly graded (uniform particle sizes)	
S	sand	W	well graded (range of particle sizes)	
M	silt	H	high plasticity	
C	clay	L	low plasticity	
O	organic			

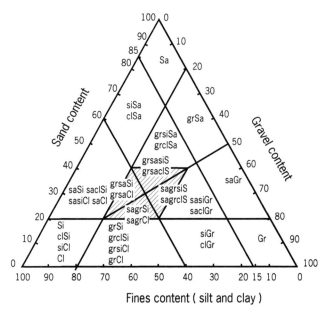

Figure 15.1 Classification of soils by grading alone.

Table 15.4 Example of extension of USCS into engineering properties

Group symbol	Characteristics for embankments	Characteristics for compaction	Characteristics for foundations
GW	Very stable	Good. Tractor rubber wheeled or drum roller	Good bearing value
GP	Reasonably stable	Good. Tractor rubber wheeled or drum roller	Good bearing value
GM	Reasonably stable	Good. Rubber wheeled or sheepsfoot roller	Good bearing value
GC	Fairly stable	Fair. Rubber wheeled or sheepsfoot roller	Good bearing value
SW	Very stable	Good. Tractor	Good bearing value
SP	Reasonably stable	Good. Tractor	Good to poor bearing value
SM	Fairly stable	Good. Rubber wheeled or sheepsfoot roller	Good to poor bearing value depending on density
SC	Fairly stable	Fair. Rubber wheeled or sheepsfoot roller	Good to poor bearing value depending on density
ML	Poor stability	Good to poor. Rubber wheeled or sheepsfoot roller	Very poor; susceptible to liquefaction
CL	Stable	Fair to poor. Rubber wheeled or sheepsfoot roller	Good to poor bearing value
OL	Not suitable	Fair to poor. Sheepsfoot roller	Fair to poor bearing value. May have excessive settlements.
MH	Poor stability	Poor to very poor. Sheepsfoot roller	Poor bearing value
CH	Fair stability	Fair to poor. Sheepsfoot roller	Fair to poor bearing value
OH	Not suitable	Poor to very poor. Sheepsfoot roller	Very poor bearing value
Pt	Not suitable	Compaction not practical	Remove from foundations

approach geologically, there is little input to provide information on the engineering characteristics of the soil.

It is interesting to note in passing that this approach puts the boundary between fine and coarse soils at 40% fines. This is in contrast to the USCS division at 50% fines. The descriptive approach advocated in this book takes this boundary to be at about 35% (depending on the overall shape of the soil grading curve) based on the behaviour of the soil. These may appear to be quite similar to each other, but in reality there remain large differences between these definitions of the coarse–fine soil boundary, see Chapter 4.7. The significance of these divisions is one reason why this definition is not given in EN ISO 14688-2 as agreement was not possible.

15.3 Classification systems taking account of engineering properties

The shortfall in all of the above classification systems is the limited amount of information they provide on the engineering behaviour of the soil. One approach that achieves this is that adopted in UK practice for classification of materials for use in earthworks for roads (Highways Agency, 2007b). This approach takes into account grading and plasticity, but also includes water content, strength of particles,

Table 15.5 Summary of earthworks classification including wider range of parameters

Class	Description	Typical use	Parameters used in classification[1]						
			Grading	Uniformity	Plasticity[2]	Water Content	Strength[3]	Chemical[4]	Other
1A	Well graded granular	General fill	*	*			P		
1B	Uniform granular	General fill	*	*					
1C	Coarse granular	General fill	*	*					
2A	Wet cohesive	General fill	*		PL		Ur		
2B	Dry cohesive	General fill	*		PL	*	Ur		
2C	Stony cohesive	General fill	*		PL	*	Ur		
2D	Silty cohesive	General fill	*			*	Ur		
4	Various	Landscape fill	*			*			
5	Topsoil	Top-soiling	*						
6A	Selected well graded granular	Below water	*	*	PI		P		
6B	Selected coarse granular	Starter layer	*		PI		P		
6C,D	Selected uniform granular	Starter layer	*	*	PI	*	P		
6E	Selected granular	Capping by stabilisation	*		LL, PI			*	
6F	Selected granular	Capping	*			*	P		
6G	Selected granular	Gabion filling	*				P		
6H	Selected granular	Drainage layer	*		PI		P	*	Friction/adhesion
6I	Selected well graded granular	Reinforced earth fill	*				E	*	Friction/adhesion
6J	Selected uniform granular	Reinforced earth fill	*				E	*	
6K,L	Selected granular	Bedding, buried steel	*		PI		P	*	
6M	Selected granular	Surround, buried steel	*		PI		P	*	Permeability
6N	Selected well graded granular	Fill to structures	*				U, E, P	*	
6P	Selected granular	Fill to structures	*			*	U, E, P	*	Permeability
7A	Selected cohesive	Fill to structures	*		LL, PI	*	U, E	*	
7B	Conditioned PFA	Fill to structures	*			*	U, E	*	Density, Permeability
7C	Selected wet cohesive	Fill to reinforced earth	*		LL, PI	*	E	*	Friction/adhesion
7D	Selected stony cohesive	Fill to reinforced earth	*		LL, PI	*	E	*	Friction/adhesion
7E	Selected cohesive	Lime stabilisation	*		PI	*		*	
7F	Selected silty cohesive	Cement stabilisation	*	*	LL, PI	*		*	

[1] Guidelines for some parameters are provided as indicated by asterisk.
[2] Plasticity: PL = plastic limit, LL = liquid limit, PI = plasticity index.
[3] Strength: Ur = undrained shear strength of remoulded soils, U = undrained shear strength, E = effective shear strength, P = particle.
[4] Chemical includes pH, sulphate content, chloride content, organic content, resistivity, redox potential, microbial index activity.

chemical considerations and permeability. A condensed summary of this classification is presented in Table 15.5.

15.4 Rock classification and rating schemes

The approach to the classification of rocks has both differences and similarities to the classification of soils. The similarity is that the purpose of the classification is to connect the present project into precedent experience through the assumption that materials of a similar class will behave in a similar way. In soils this is done through classifications that are defined on the basis of the material characteristics; there are few if any classifications of the mass characteristics of soils. In rocks the mass features such as discontinuities are so much more important and rock mass classifications are widely used. As these cannot easily be measured by testing, the emphasis is on mapping the conditions at one site onto the conditions encountered on sites previously. The classification therefore acts as a vehicle for the designer to benefit from previous experience in terms of the excavatability, overall rock conditions and support requirements which should be anticipated on the site in question.

Early rock classification schemes used few parameters, were entirely empirical and mostly related to underground excavations. The earliest rock classification system was proposed by Terzaghi (1946) and is important here because the simplicity was supported by the expressly stated importance of the geological investigation as an early part of the design process. Later classifications were proposed by Stini (1950), Lauffer (1958) and Deere and Deere (1988). As the empirical database on which these classifications were based grew in size and scope, the inclusion of additional parameters became relevant leading to the Council for Scientific and Industrial Research (CSIR) (Australia) system (Bieniawski, 1976) and the Q system (Barton *et al.*, 1974). There has been considerable further evolution of these schemes into related areas such as tunnelling machine performance. An important element of note is that the only rock material parameter within these classifications is the material strength. All the other parameters relate to the joints (including orientation, spacing, strength, persistence) and the in situ stress and groundwater conditions.

One of the major problems confronting designers of engineering structures in rock is that of estimating the strength of the rock mass. A rock mass consists of interlocking discrete blocks. These blocks may have been weathered to varying degrees and their boundary surfaces may vary from clean and fresh to clay filled and striated.

Determination of the strength of an in situ rock mass by laboratory type testing is generally not practical and it becomes necessary to estimate the strength from the description of the rock which is made up of geological observations and from tests on individual specimens of the rock material or discontinuity surfaces. This question has been discussed extensively by Hoek and Brown (1980) who used the results of theoretical studies (Hoek, 1968), work on models (Brown, 1970; Ladanyi and Archambault, 1970) and the available strength data, to develop an empirical failure criterion for jointed rock masses. Hoek (1983) went on to propose that rock mass classification systems, such as those mentioned above, can be used as a tool to help in the estimation of the rock mass constants required for the empirical failure criterion proposed. The range and history of rock mass classification schemes in engineering practice are summarised by Bieniawski (1989).

These rock mass classification systems have become very useful and practical tools, not only because they provide a starting point for the design of the works but also because they force users to examine the properties of the rock mass in a systematic manner. Putting this another way, the rock has to be examined and described in a systematic manner to allow the design process to set off in the right direction.

These approaches involve accurate and comprehensive field description allied to measurement of strength for correlation with previous experience. The description of the rock mass should be carried out as outlined in this book. The use of rock mass classification systems is beyond the scope of this book and the reader is referred to relevant texts such as those cited above.

16 The Description Process – Boreholes

Description is both a science and an art which requires that a methodical process is followed in order for there to be any chance of a good description being the outcome. This description is incorporated into the product of the whole process which is a set of internally consistent verified data initially put out in hard copy form as a field record such as a borehole or trial pit log.

The need for a systematic approach to the description of soils and rocks has been stated earlier. This approach includes the word order and the use of both the defined terminology and good English. However, a good description is the result not of a haphazard series of actions but rather the result of an ordered approach to the whole business of description. The ordered approach necessary to achieve these aims starts with the preparation for field logging, follows with the opening of the soil samples or core boxes on arrival at the sample store and continues right through to the completed field log.

This chapter describes the steps that need to be taken in the process of soil and rock description, in order to achieve the stated objective. In principle, the approach is the same whether the description is being made on rock cores or discontinuous samples in the core store, or on exposures (natural or created) in the field. However, there are differences in the practicalities and these two sources of exposure for description are covered separately. This chapter deals with the logging of borehole samples and Chapter 17 outlines the steps towards logging field exposures.

The first point to remember in compiling a description is the definition of a borehole log. In Clause 47.2.6.1, BS 5930 explicitly defines the borehole log as 'a record that is as objective as possible of the ground conditions at the borehole position before the ground was subjected to disturbance and loss by the boring process'. It is clear that some interpretation will almost always be required in order to compile a field borehole or exposure log. The amount of interpretation that is required depends on the nature of the sample or exposure available for description as shown in Table 16.1.

The point is that description of cores or samples is not just a description of the actual core in the box, nor assessing the broad character of the ground mass, but is as objective a view as possible of the ground in situ at the borehole location before it was disturbed.

The initial field description on the sample or core description sheet or trial pit log can be a factual record of what is recovered. Indeed it is good practice to include all potentially relevant information on this primary record. However, the interpretation also needs to start at this point, as there are decisions which are best made when the material is still available to the logger. This includes assessment of what elements of the presented material are artefacts of the creation of the presentation. This is the point at which the logger needs to decide whether the soil or rock has been softened, mixed, fractured or lost as a result of the drilling or excavation

Table 16.1 Extent of interpretation in description

Nature of sample	Amount of disturbance	Remark
Description of natural exposure	Amount of disturbance should be minimal although weathering (stress relief and water content changes) might need to be considered.	
Description of excavated exposure	Amount of disturbance depends on nature of ground and method of excavation. Disturbance could be negligible.	The amount of interpretation required to arrive at a field description generally increases as the amount of disturbance in the creation of the exposure increases and as the completeness of the exposure available decreases.
Description of continuous core	Amount of disturbance depends on nature of ground and suitability of drilling equipment for the ground, but will include some disturbance from core extrusion and handling.	
Description of discontinuous samples, generally from some form of soft ground boring method	Amount of disturbance likely to be significant in some samples and the interval between samples has to be interpolated and completed from other observations, such as drillers log or other logging records.	
Compile stratum description	Having compiled individual descriptions and observations a stratum description requires that certain information is edited out and the description 'averaged' for the range of materials seen.	

process. These decisions are sometimes difficult or worse, as discussed below, but these decisions are even more difficult after the logger has left the logging shed or exposure. For this reason, the logger also needs to compile the final descriptions at this time.

It is a sensible requirement that the amount of interpretation within a field description needs to be kept to a practical minimum. Sensible assessment of the actual ground conditions in situ is required in order to comply with the above rule. On the other hand, speculation is not helpful to anybody and should not be entertained. The approach and steps that need to be followed to achieve this balance are covered in the following sections.

16.1 The approach to description

Logging soil or rock samples or cores is physically hard work. It can also be a fascinating art, an exacting science or very boring. Those who are bored by the process have not prepared properly for the task in hand. In order to make a good job of description, the logger needs to know what geological conditions are expected and why the investigation is being carried out. They need to have a reasonable idea of the ground model at this stage and where the uncertainties are within the model. Without this information deviations from the expected conditions may not be identified and the information recorded on the field log may not be entirely appropriate to the engineering problem for which the investigation is being carried out. In other words, the ground model will not be tested and updated as the investigation progresses.

The logger should ensure that they are aware of this information, either through a proper briefing or through their own reading of the project file. This can be enhanced by a desk study, which could be as little as a review of the geological map of the site area together with looking at copies of previous boreholes in the locality. This information will be of substantial benefit when the logger starts reviewing the ground conditions, as shown by the samples and cores being examined.

Once in the field and carrying out the logging, the logger will benefit from compiling cross sections of the borehole records. Even the simplest of stick cross sections will be sufficient in most cases to provide a running check on the succession and the ground model, with the various strata identified by simple colour coding. The benefit of this approach is primarily to check the model and the assignation of formation names to the strata units for consistency and sensibility. Again this sort of check is best carried out live as part of the description process when sample descriptions can easily be reviewed and samples re-examined if necessary. This is so much simpler than wondering later why the strata do not fit and having to go back to the field store and find the samples, which by then will probably be well buried. There is an added but hidden benefit in that the logger who is carrying out this sort of check will be more interested in the ground conditions and the descriptions being prepared and will remain interested for a much longer time that the logger who does not engage with the task in this manner.

The logger should remember at all times that they are carrying out an information gathering role which comprises an important element in the whole investigation. As such, the logger has certain areas for which he or she bears primary responsibility. These can be summarised as recording the materials encountered in the investigation and ensuring that the compilation of the exploratory hole records is fully and correctly carried out. The work that goes into meeting these responsibilities is described below.

16.2 Logging equipment and the toolbox

Good logging needs, in addition to the methodical and ordered approach, good working conditions. These are both to enable good descriptions and thus a high quality field log to be compiled, but also to enable the logger to keep going on this activity all day and the next day. A cold draughty ill-lit shed where the papers get wet is not conducive to such an outcome.

The appropriate conditions that are required include:

- General weatherproofing of the logging area so that it is reasonably warm (or cool) and dry. This can often be most easily achieved by placing some sort of cabin within another building which is the sample store. If the logger is kept at a sensible temperature and provided with the means to make appropriate cold or hot drinks, the productive day is longer and the quality of work within that day is also markedly higher.
- Good working conditions include good light. The light should preferably be natural light. Working in strong direct sunlight is not ideal but a bright cloudy

day is perfect. Arranging the logging area so that the working area is in the shade will suffice in some circumstances. In practice most logging areas are indoors and artificial lighting will have to be provided. This should be satisfactory particularly if logging benches are arranged near large windows on the shady side of the building. If lighting needs to be provided, daylight (CIE D65 or CIE C) fluorescent tubes should be used. In any circumstances, the use of colour charts will help alleviate problems associated with variable lighting.

- Benches to lift samples or core off the floor are preferable. Most sample description areas on site do not have this facility and the loggers spend the day crawling around on the floor. Whilst this is perfectly feasible, it is not ideal and the best logs, in my experience, come when the logger can stand up (or perch) and work at a bench. The downside of providing benches is the need to lift samples and core boxes onto and off the bench, thus requiring additional help to do so. This is not difficult to organise and still provides a more efficient and better quality product.
- A supply of water is needed to enable wetting of core or samples, execution of the hand tests for fine soils and to keep your hands clean.
- In addition, it is necessary to have a supply of potable water to avoid dehydration when working out in the open air during a long day.

Logging is a physical activity which requires the logger to have a tool box of equipment to enable the process. This will include paper guidance and record sheets and a range of logging and testing items. In addition, the core store logger will need equipment to assist in taking subsamples, resealing bags, tubes and boxes. The contents of a basic tool box are listed in Table 16.2 and examples shown in Figure 16.1.

Table 16.2 Basic logging tool box contents

Paperwork	Logging tools	Testing tools	General tools
• Logging manuals • Standards and specifications • Material and formation-specific logging guidance • Proformas for sample and core description • Testing record sheets • Drillers records	• Gloves • Tape measures (3 or 5 m and 30 m) • Water spray • Bucket of water • Hand lens • Pocket knife • Scrapers/spatulas in a range of sizes • Geological hammer • Protractor or clinometer • Pencils and erasers	• 10% acid (HCl) • Hand vane • (Penetrometer) • Point load tester • Schmidt hammer	• Sample trays/washing up bowls • Sample bags or tubs • Bag ties • Adhesive tape • Labels and marker pens • Claw hammer and nails • Paraffin wax • Wax kettle • Timber to replace core samples
	Photography items	• Camera • Colour charts • Gray scale charts • Photo header board with scale	

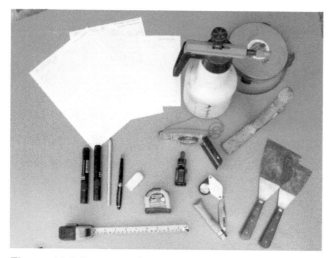

Figure 16.1 Basic set of some of the most essential logging tools as listed in Table 16.2.

However, every logger has their personal preferences and adjusts the contents of their own box to their preferences. Very few of these items are not mandatory.

Clearly there is a large number of items here, some of them bulky and these are not readily carried around by one person. Nowadays most loggers tend to have a vehicle to get to and from the site and this enables the tool box to be transported to the site. It may then need to remain onsite for the duration of the project.

There is a problem when loggers are required to make a short visit to a site. Nowadays this may just be a day visit to a site for some check logging or to inspect some samples. If travel on such visits is made by road or rail, an appropriate reduced tool box can be transported without undue difficulty. If however, travel is by air, there is clearly an issue with taking tool box items on the aeroplane other than as hold luggage. This is perfectly feasible but shortens the amount of time available on site after waiting for the hold bags to come to the collection area in the airport. Alternative arrangements for the basic tools to be provided locally may be of assistance on such occasions.

16.3 Description of samples and cores

A suitable area to carry out the logging is established and the support and tools that are necessary are assembled. The next step is to receive the samples or cores and to begin the actual logging.

The first step in the methodical approach is to lay out of all the samples or core boxes in depth order, either from the top down or from the bottom up as shown in Figure 16.2. Ideally this includes all the samples for an entire borehole, although this is not always feasible for reasons of space, delivery of samples or possibly for contract requirements. In any case, the samples laid out should comprise a significant proportion of a borehole and certainly not less than about 20-m worth, unless the hole is shorter than this, of course.

A systematic reconciliation of the sample numbers and depths, run depths and recoveries should be made against the lead driller's record sheets to ensure internal

Figure 16.2 Laying out all the samples before starting the logging of a borehole (photograph courtesy of Soil Mechanics).

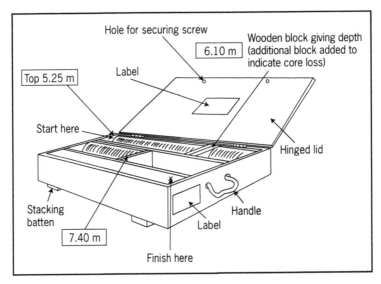

Figure 16.3 Convention for construction and labelling of core boxes.

consistency of all information. Any discrepancies should be resolved at the first opportunity and preferably before logging is carried out. This will usually necessitate a conversation with the driller.

The conventional layout of a core box is shown in Figure 16.3. The core is placed and read as in a book, starting at the top left where the top is the line of the hinged lid. The runs are placed in consecutive order and separated by core blocks with the depths marked. If the cores have to be cut to fit, this should be noted on the box. Labels should be placed inside the lid, outside the lid and on the end of the core box.

The log should include a record of the sample recovery as a length, such as the length in a tube sample or the length of a core run (see Chapter 12). This is best

---

TIP ON FIRST STEPS IN CORE LOGGING

- The initial stage of the core logging which is of critical importance in getting things right includes review of the total core recoveries in order to make any depth adjustments that turn out to be necessary.
- Depth adjustments can be necessary, either because core was dropped and then recovered in the next run, or because there is core loss and the zones of loss need to be assigned. In either instance, corrections will need to be made to the depths of any features in the core, whether these are stratum boundaries, changes in fracture state or the location of subsamples or tests.
- This initial stage of assessing and tidying the core is invaluable as a preliminary description stage. The logger will literally be getting a feel for the materials, their variability and considering some initial decisions on the strata boundaries.

---

carried out at a very early stage to permit any corrections to depth that arise from variable recoveries before any other records are made and certainly before any subsamples are ascribed depths. In addition, the record of the sample length, together with a note on the quality if derogated in any way, will be of great assistance to the person scheduling the laboratory testing.

The samples should then be individually described in accordance with the procedures given in Chapters 4 to 14. Initially the samples are described individually and systematically, using the defined word order, essentially factually. The consistency of descriptions by one individual and between other logging individuals working on the same project is greatly assisted by using aids such as colour charts and reference samples. The establishment of these aids for team consistency is a matter for the individuals within the logging team on site.

In principle every sample should be examined and a description recorded. The only exception that should normally be permitted is when the shoe from a tube sample is available and representative, inspection of the base of the tube may not add to the record, particularly as the shoe is usually the least disturbed material. Leaving the tube contents intact may be useful in laboratory testing of the specimens.

TIP ON SEQUENCE OF BOREHOLE LOGGING

- It is normal practice to describe boreholes from the top down. There is no reason for this except that this is the direction of drilling, starting at Sample 1 or Run 1.
- However, a large majority of variability in the ground owing to weathering, other geological processes or man's interference is usually within the upper few metres so the logger must tackle this variability at the start of the logging exercise.
- Consideration should therefore be given to logging from the bottom up. The variability is less at first and when a zone of marked variation is reached the logger is already aware of the characteristics of the succession.
- This approach is commended as sound practice.
- Once the logger has 'got their eye in' for the succession at a site, the room for confusion in the upper metres is reduced and practice can revert to logging from the top down.
- In addition, the position of marker bands is usually easier to identify in relation to the base of the unit.

Descriptions and all other observations or notes should be recorded on a description sheet. Sample or core description should always be carried out using standard proformae issued by the contractor. This is to ensure that the records find their way into the project file and to make use of the prompts for information that are required by the Standards and which should be included on the form.

At the very least, if for any reason a field notebook is used (such as in wet conditions), it should be mandatory for the information to be transferred to a proforma at the end of the day and signed off by the logger. If the company is ever in the position of having to defend a field log in a contractual argument, and this will usually be several years later, the absence of these primary field records will seriously hamper any defence.

Example proformae for field use are included in the Appendix. Copies of the proformas are also available on the author's web site for download and use at www.drnorbury.co.uk.

The sample or core description sheet is the opportunity for the logger to record all information about the samples or cores that may or may not be relevant to a complete stratum description. It is good practice to record even negative information, as this forms a permanent record that the logger has thought about that aspect of the description and made an observation, for example, to record that the 'gravel is fine to coarse', even though the final stratum description may only state 'GRAVEL'. Anyone checking the log will be content that the grading of the gravel has been considered. Similarly, the absence of fissuring in a stiff clay is a useful observation in its own right, as it provides information on the history and grading of that particular clay.

Having described the samples, the stratum boundaries need to be identified. This is best done whilst all samples are open and available at the time of logging. The boundary should be found by standing back and looking at the samples, taking guidance from the driller's log and marking a boundary depth on the sample

---

TIP ON WHAT TO RECORD ON LOGGING SHEET

- It is good practice to use the sample, core or trial pit logging sheet to record ALL information and observations. Excess information can be edited out later, but it is useful to have a record of all tests carried out, all variations and extremes of the material consistency, grading or lithologies.
- Examples of information that can be recorded and which will generally not appear on the field log include:
  - consistency on disturbed samples
  - full range of discontinuity spacing
  - the presence of rare minor constituents which may or may not be significant
  - grading of the sand and gravel, even if fine to coarse
  - the presence of drilling-induced disturbances
  - samples which are considered to be unrepresentative
  - samples which are too small to test
  - carbonate free status when there is no reaction to acid.
- The logger should bear in mind that if there are any questions about the logged information, it will always be easier to review a field sheet, with or without the aid of photographs, than to have to go and find the samples or cores in the store.

or core description sheet. The logger should never ignore or change the foreman driller's boundaries or depths without very good reason. He saw all the soil, not just the samples presented and there must be strong positive reasons for any such amendment. Additional boundaries that the driller did not record on his log may, of course, be added.

Compilation of the stratum descriptions can also sensibly be made at this point with the samples still open or the compilation could alternatively be left until later when a seat is available; there is no strong preference either way. However, the logger should not progress to the next stage of closing up the samples or boxes and placing them in storage until they are happy that all necessary information has been extracted from the samples and cores presented. The guiding rule here should be one of common sense. It is much easier to record all information once, rather than have to come back to the store and find the samples again on a subsequent occasion.

In editing sample descriptions to arrive at stratum descriptions and in completing the field log, the ground model needs to be considered and tested. If there are features present in the soil that are critical to the interpretation of the geological succession, these should be included. For example, the lithologies that identify which glacial till is present or the single brick fragment that identifies made ground are usually significant and important. The presence of a few unnamed shell fragments or a few grains of sand in clay are much less likely to be critical.

When carrying out the description in the field, the logger should have in mind the appearance of the completed borehole log. There is no point in spending a lot of time recording information in minute detail if this is not merited by the needs of the investigation. The reason why most field logs are presented at a scale of 1:50, which is equivalent to 10 m per page, is that this amount of information is normally found to be sufficient for most projects, with enough space to report the most critical and interesting features observed in the samples or cores. It is suggested that deviating from this scale should only be through a positive decision made by the project leader, not simply a default because the logger has recorded too much information. This option arises too often in today's world where borehole log programs often make these decisions without any control, a classic case of the tail wagging the dog.

The use of shorthand abbreviations on field logging records is perfectly permissible and reasonable and can make the logger's life much easier. Writing the descriptions out in full is always acceptable and could be considered best practice. However, in a large investigation, or uniform geology, this can entail a lot of effort for little gain. Whether or not to use abbreviations depends how much logging there is to be done and who is going to transcribe these onto the finished log. If there is any doubt at all, descriptions should be written out in full.

TIP ON EDITING

- The compilation of a stratum description will require that some information recorded on individual samples needs to be edited out. This is to make the final description concise, to the point and readable. The decision on what to include and omit is largely one of common sense. Leaving out the extremes of grading, the full range of gravel lithologies or every variation in joint spacing may be reasonable.
- However, it is important to remember always to retain information that is diagnostic, however minor or rare that constituent is in proportional terms.

TIP ON THE SCALE OF LOGGING

On the basis of logging at 10 m per page, the logger should mentally be logging at this scale and thus will find that:

- Main descriptions should normally represent no less than 1 m (normally use detail indent for any thinner sub-units) and no more than 20 m thick. There are few geological units in the engineering zone which are that consistent.
- A stratum should not normally include more than one change with depth such as 'becoming stiff below 5 m' or 'becoming sandy from 6.5 m', otherwise it can become difficult to decide how much of the original main description still applies. It is better to start a new stratum if there are more than, say, two gradational changes.
- Zones with uniform fracture spacing in rock should generally be between 0.5 m and 5 m thick.

TIP ON THE USE OF SHORTHAND ABBREVIATIONS

- Any shorthand used must be consistent and intelligible to any other readers. Personal codes are not to be used under any circumstances. The system used must be readily understood by other loggers, technicians carrying out data entry and, heaven forbid, lawyers if things go wrong later.
- If any shorthand notation is ambiguous, more letters must be used; the example of bl meaning either blue or black below is a good one.
- Examples of acceptable shorthand:
  - gry brn fm occ c GRAVEL
  - stiff ex cl fiss bl grey mott or brn CLAY (NOTE bl is not good – use blu or blk)
  - wk brn fg SDSTE
  - mod str blk x cl jted MDST
  - v str mg white GRANITE
- A longer list of acceptable abbreviations is given in Appendix 1 of Blackbourn (2009).

There is no rule that says shorthand abbreviations have to be used and if there is any possible doubt about clarity or possible ambiguity, shorthand should not be used.

16.4 Check logging

Most logging is carried out by the younger members of our profession, but the level of supervision and checking of the logging by these junior staff is variable. This seems strange as the liability of the investigation company is resting substantially on the logger's young shoulders. Graduate logging staff may or may not have received training in systematic logging for engineering purposes. Few companies provide formal training in systematic soil and rock description such as is described in this book. However, most companies do provide some level of mentoring and support, although this can be minimal. This sort of support is critically important given the inexperience of many loggers.

Check logging is the main practical element of this mentoring and comprises two different activities: check logging in the field and log checking, which is usually a later, office-based, activity.

In the early days of a graduate's practice, the activity of check logging comprises review of the description by a more experienced individual. The check logger has to go out to the trial pit or core shed and look at the samples themselves, preferably in the company of the logger. All aspects of the sample and stratum descriptions that are presented on the preliminary field records should be checked. The transfer of information to the field log should also be checked. This procedure can progressively be relaxed as the graduate gains experience and demonstrates the necessary ability. However, each time the graduate goes into a new set of geological conditions the procedure has to be reviewed so that an appropriate level of checking can be carried out. A step back to a closer level of checking could well be necessary.

Check logging is an activity that includes substantial elements of teaching and mentoring, as well as being a measure to control corporate liability. The check logger is responsible for helping the logger to ensure that the materials encountered in the investigation have been fully and correctly recorded by the company.

The subsequent activity of log checking is where a senior member of staff or delegated person within the project team reviews the completed draft field logs individually and as a complete set. The logs can then be signed off by a senior member of staff in accordance with the necessary peer review or quality assurance procedures on behalf of the company. This senior member of staff is the responsible expert in accordance with EN ISO/TS 22475-2. It is a requirement of the Standard that the responsible expert is satisfied that the descriptions are correct, that the logger and check logger are appropriately skilled and qualified, that the information on the log is correctly compiled, that the logger's thumb or hammer swing is suitably and correctly calibrated against the descriptive terms applied, that there is consistency in approach between the various loggers and that the logs fit the geological model.

16.5 Photography of samples and cores

Photography of the samples or cores forms an appropriate part of the permanent record of the recovered samples after the samples or cores have degraded with time or been discarded. Photography of soil samples is not routinely called for in investigation specifications, whereas photography of rotary cores is normally a contractual requirement, although this only became a routine requirement in the late 1980s.

The traditional limitation on what was called for in terms of photography arose from the use of cameras with film and this resulted in the not inconsiderable time and expense of taking, processing and printing photographs on paper. With the current availability of digital cameras, there is no longer any reason not to take a comprehensive set of photographs of recovered materials, both overall of larger scale features in the soil or rock mass and also in close up of any features of interest, such as those listed below. It is a matter of communication and common sense as to the best way to record and transmit these images as elements of the investigation data, but it is not necessary to include printed copies of all photographs taken in final reports. A cross reference to an image file is often preferable.

Photographic images can be taken of the following:

- rock cores usually before disturbance caused by logging, sampling and testing
- cores of brickwork and macadam in much the same way as for rock although often taken after logging

- soil cores as for rock core but which may be split before photography
- extruded tube samples before and after splitting
- laboratory test specimens after testing
- trial pits and other excavations including excavated spoil
- field exposures
- features of interest such as marker beds, fossils or unusual features within the soil samples or rock cores taken in close up.

Apart from the last of these, the photographic record is a systematic exercise with photos being taken of every sample or core, as required for the project. Taking photographs of features of interest is clearly an *ad hoc* exercise.

Photography can be appropriate at several stages to illustrate changes or particular features of the soil. For example, there are occasions when air drying of the core reveals fabric features which are also of interest. This is most likely to be the case in fine soils where features such as laminae, partings, fissures and shears are only clearly visible after a day or two of drying. If this is the case, photographs of these features can usefully support and enhance the descriptions on the field logs.

All photographs should include:

- a header board with details of the project, sample identifier including borehole and sample numbers and depth reference and date of image. The header board should also include some sort of scale which is usually marked in 10 cm bars. A tape measure should be included in any close up pictures. Note that the project name should be identified at the very start of the investigation as it is not possible to change such details later. Common practice today is to print off labels from the site computer which provides tidy labels, but the black on white arrangement is far from ideal; header information is much better on a grey background.
- photographic colour charts with standard ranges of colours and a chart of grey tones. These colour charts are readily accessible from suppliers on the internet.

There are a number of points that also need to be considered in taking photographs of soil and rock samples:

- Header boards and backing boards should be painted mid grey. Use of a matt undercoat is acceptable as it provides little reflection.
- Care should be taken in all photographs that the plane of the core, the boards and the focal plane of the camera are all parallel, as shown in Figure 16.4. Any non-parallelism will increase the distortion of the image as shown in the poorly arranged photograph of core in Figure 16.6.
- Lighting considerations are the same as those identified above for good logging conditions and colour description. The optimum position for photographs is in the entrance of the sample store or outside, preferably on a bright but cloudy day. Full sunlight should be avoided as it creates strong shadows which effectively mask the core. Finding the right camera settings to adopt is a matter of trial and error, but generally setting the camera to an automatic option will produce the best results. A little bit of trial and error will demonstrate whether the photographs will be best with or without help from the camera's flash or other lighting arrangements. Additional lighting is required much less often today as the capability of camera flashes has improved.

Examples of core, pit and exposure photographs are presented in Figures 16.5–16.8.

Today, when digital photographs can be edited in a report, there is no reason why the photographs cannot be digitally annotated to show depths, samples or other features.

The traditional requirement for core photography (of soil, rock or made ground) is for an image of a complete box, which is fine as a record of the overall core recovery in terms of the amount and condition of the recovered core. The requirements for this image are normally specified and the photograph should be taken before there is any disturbance caused by the logging procedures. However, because of the shape of core boxes in relation to the proportions of normal photographic images, most of these images include a lot of header board and not much core. Every effort should be made to maximise the proportion of the image that is core. Nevertheless, the detail that is available on the core is usually limited. To counter this problem, the resolution available with better modern digital cameras allows printing of images up to A4 size, so that much closer examination of the core and features within the visible profile of the core is possible.

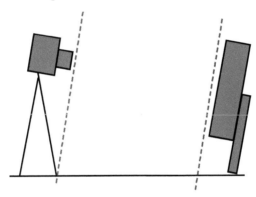

Figure 16.4 Principle of photography set up with camera and image co-planar.

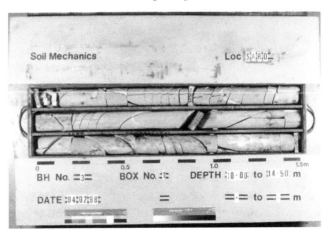

Figure 16.5 Rock cores (artificial): example of reasonable balance of photograph.

Figure 16.6 Example of informal core photograph without header boards. Note the distortion in the photograph caused by the inclination of the camera with respect to the cores.

Figure 16.7 Trial pit record photo with good lighting and scale and visibility to the base of the pit (photograph courtesy of Soil Mechanics).

Figure 16.8 Photograph of exposure, complete with scale, although lacking any location information (photograph courtesy of Chris Eccles).

Figure 16.9 Core scanner for high resolution imaging of cores, around whole circumference if core is sufficiently intact (photograph courtesy of DMT GmbH & Co. KG).

Close up imaging of sections of single core channels provides an alternative approach. The high resolution images allow off-site close examination of all aspects of the recovered cores, including the lithologies and the fractures. These images can be obtained either by laser scanning or using a camera.

An example of a portable laser scanning device is given in Figure 16.9. Images of full circumference or slabbed core or complete core boxes can be recorded. Full circumference core is scanned at a rate of approximately 0.6 m per minute and at a standard resolution (150 dpi) for quantitative fracture evaluation or at very high resolution (up to 1000 dpi) to enable image analysis such as grain size determination. Core images are stored in a digital drill core library which can be made accessible to team members off site.

An aspect of particular value is that the scanning can in theory be carried out prior to transport, thereby eliminating one aspect of damage to the core. The scanned image can be analysed to produce geotechnical parameters, such as rock quality designation (RQD), joint frequency or joint spacing. In addition, assessment can be made of the petrographic properties.

Core scanner images can be at a sufficiently high resolution to allow detailed examination of the cores and to form a worthwhile permanent record of the recovered core. However, the equipment is expensive and the core handling and image processing efforts are labour intensive so this approach is unlikely to be used on all investigations. Alternatively the equipment includes a facility for rotating the core to allow recording of discontinuities around the full circumference and thus opening out the trace onto a borehole log. These recording and presentational enhancements provide excellent information and can be useful, but the rock to be scanned needs to be of high quality in order to allow the individual pieces to be placed onto the scanner rollers. In practice, many engineering geological investigation cores are not in rock of sufficient quality and the core must be placed onto the scanner while still in the core liner. Under these circumstances, the scanning product is simply a high resolution image of the top surface or a cut surface of the core.

The key advantages of this sort of approach include obtaining high resolution images from a constant distance and with a constant 'exposure' and thus with a consistent scale and colour. This allows objective viewing of the images throughout the formations within the succession being investigated.

All photographs should be taken with the sample or core damp. This may require the core to be moistened with a fine water spray before taking the photograph. Care should be taken to avoid excess moisture, as the film of water will reflect light and obscure features on the core.

16.6 Testing and sampling

Having completed the description and photographed the soil and rock samples and core, the final activity is to carry out any field testing on the samples or cores that is scheduled and to take any samples for preservation. Samples are frequently required to be taken from the cores immediately on recovery to the surface. The comments below apply equally to the samples taken at that early stage. If at all possible, description of the sampled material should be made before the samples are removed for preservation and subsequent transport to the laboratory.

Field testing on recovered samples can comprise a range of tests. These can include hand vane, point load, Schmidt hammer, water content and density determination through hammer pick and nail tests. The test procedures are not covered here; instead the reader is referred to the relevant Standards for test procedures. These will be in various sources such as EN ISO 22476, BS 1377 in ISRM suggested methods (including point load and Schmidt hammer) or in ASTM or other national publications. All test results should be recorded in the field using the appropriate test record sheets. Testing should also be carried out as appropriate to check the calibration of the logger in regard to strength.

Subsamples should be taken from tube, bag or core samples as appropriate and testing scheduled. Care should be taken to transfer the sample details to the new container, including way up indicators on core samples. Testable samples of core should be removed from the core box ensuring that the length is at least 2.5 diameters and preferably three times the diameter. The sample should be wrapped to provide preservation and protection in accordance with normal local practice. This typically includes layers of cling film, foil and wax, with the complete wrapped sample being placed in guttering or core boxes for transport. The preserved samples should include identification labels with project, borehole, depth and orientation markers placed both on the inside and the outside of the layers of protection.

It is not generally preferable to leave core inside the semi-rigid core liner that is widely used today as this has too much clearance to restrain the sample adequately; the suitability of this liner for preservation will depend on the material and the testing that is proposed.

After completing the sample or core description the samples should be prepared for storage in accordance with the following check list:

- Any debris from the logging should be cleaned off or removed.
- Samples should be closed up and resealed and the sample labels refreshed as necessary so that they are visible and legible.
- Any samples containing hazardous material such as glass, wire or chemical contaminants should be labelled as such.
- Tube samples should be re-waxed.
- Cores should be tidied and secured in the boxes.
- Core liners should be replaced and sealed. Core and the liner can usefully be wrapped in cling film to preserve moisture but the decision as to whether to do this is on an individual project basis.
- Sample positions should be marked with rigid replacements such as wooden stakes, dowel or plastic pipe.
- Zones of destructive logging should be identified on the core box.

16.7 Compilation of field log

The description of the individual samples is now complete and the boundaries between strata and their depth assessed during the logging have been incorporated. It is now necessary to compile the field log and include the stratum descriptions on the log. The stratum descriptions are the element of description that is required to be as objective a view as possible of the ground in situ (Clause 47.2.6.1, BS 5930).

The first step should be to compile the skeleton log, that is the log with all information from the driller but without descriptions. The skeleton log should have included on it:

- driller's stratum boundaries
- all the driller's remarks including any relevant information about the borehole such as obstructions or borehole instability
- depths and markers on samples and core runs
- details of field tests carried out and results if required from the driller
- chiselling records
- information on progress of the borehole, overnight levels, casing levels
- groundwater observations including strikes, levels at tests, levels overnight, sealing depths of sealing off water.
- information about the equipment in use such as:
  - rig type
  - casing sizes
  - drilling and boring tools in use including core barrel type and bit design

- core diameter
- flush type
- flush returns (colour and percentage)
- details of any installations in the borehole.

It is the logger's responsibility to ensure that all of the above items of information are correctly and completely transferred to the field log.

The stratum descriptions should be added to this skeleton log ensuring that the boundaries and the descriptions are consistent with all of the information from the driller. The stratum descriptions should be presented as main descriptions which convey the overall impression of the material. This can be enhanced with local more specific detail on individual occurrences or changes. The aim as always is to present the reader with a clear picture of the ground succession at the borehole position, without obscuring this with excessive wordage which conceals the information being reported.

In order to present information on the how ground was before drilling took place, the borehole descriptions need to take account of those factors which could have caused disturbance of the samples as recovered. These factors include:

- loss of fines in sands owing to inappropriate sample recovery methods
- borehole piping affecting blow counts – note that EN 1997-2 requires that data shortfalls should be reported and not ignored
- loosening of fine sand which can occur for several diameters ahead of the borehole owing to the effects of the boring tools
- softening of clays by comparing different types of samples and blow counts
- desiccation and smear of clay in trial pit walls
- presence of angular gravel or cobbles – are these cavings or larger particles broken up by the drilling tools?
- sample degradation through extended storage
- destruction of interparticle cement in a weak rock such as sandstone
- zones of core loss
- fracturing of rock by drilling (the driller often has to break up the bottom of each run to get it out of the catcher box)
- break up of core owing to stress relief
- spinning of core pieces
- effects of samples being disturbed by storage in the sun or frost
- sample damage through rough handling in transport.

Where a sample may not be representative, assessment of the in situ state before drilling can be very difficult. The logger should draw attention to this difficulty in the descriptions. The initial levels of uncertainty can be indicated by the use of 'probably' and 'possibly'. Thus, helpful but qualified information can be provided by descriptions such as 'probably stiff' or 'possibly weathered MUDSTONE'.

Where the presented sample is highly disturbed, the assessment of the in situ ground may not be easy and it will not be wise to speculate. In these cases, use of the term 'recovered as' provides an option for the logger to show that much interpretation is required to assess the in situ ground condition from the samples that have been recovered. The reasons for such poor sample quality can include the performance of the driller as well as the character of the ground. Such circumstances should not be hidden away and this is an explicit requirement in EN 1997-2. It is easier and better to own up on the log rather than trying to be too helpful and ending up in court. The fact that this is indeed the case is useful information in its own right and can be interpreted by the reader as they see fit.

There are two cases when 'recovered as' can be considered, namely at the rockhead and in zones of low core recovery (quantity or quality).

The construction of boreholes at the rockhead is often difficult owing to the increasing compactness of the ground and involves a switch from soft ground to rock hole-making methods. As the soft ground methods reach refusal, the sample quality decreases markedly. The driller has had the advantage of seeing all the material and hearing the sound of the tools in the ground and should be well aware of the in situ condition. The samples presented, however, will be disturbed and often fragmentary. Nevertheless it should be reasonable and possible to interpret what the ground is like. It is therefore sensible that the description is presented in the sequence of the interpretation first, followed by the observed factual evidence.

In the case when core recovery is low, for whatever reason, a slightly different approach needs to be taken. This usually occurs in rotary cores where neither the driller nor the loggers have seen the entire sequence. If the core recovery is greater than about 75%, interpretation of the in situ ground condition should be reasonably achievable, as shown in the examples in Figure 16.10. However, if this is not possible and this will almost certainly be the case once the core recovery is less than about 75%, the approach should be to present the factual evidence first and to

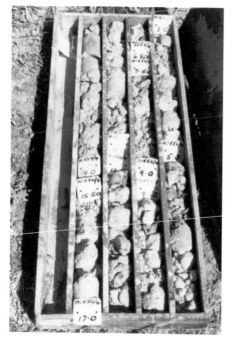

(a)

(b)

Figure 16.10 Examples of damaged core which require considerable interpretation to arrive at the assessment of the in situ condition of the rock: (a) shows a whole box of core recovered in a non-intact condition which does not represent the in situ condition, and (b) shows a close up of core damage in chalk associated with a flint band.

EXAMPLE DESCRIPTIONS

Examples of using 'probably', 'possibly' and 'recovered as':

- When the sample is disturbed:
  Probably stiff brown CLAY with extremely closely spaced partings of fine sand. Recovered as soft owing to water migrating from partings.

- Where the sample is not representative:
  Possibly soft grey slightly organic CLAY. Recovered as hard owing to desiccation before logging.

- Where sample is disturbed near boring refusal at rockhead:
  Probably weathered MUDSTONE. Recovered as clay with fine gravel.
  (In this example, the rock legend should appear on the borehole log)

- When core recovery is low, say 25–50%:
  Partial recovery. Core loss presumed to be more weathered material. Recovered core comprises medium strong grey coarse grained SANDSTONE. Sand and gravel size fragments recovered. Heavy discolouration on discontinuity surfaces, penetrating up to 3 mm. (probably weathered COAL MEASURES SANDSTONE).

follow this with the logger's best assessment of the in situ condition. Appropriate forms of words for use in these circumstances are provided in Table 12.2.

The compilation of the stratum descriptions requires selective omission of information and inclusion of relevant information from the driller and any other sources of information. The draft field log should be compared with the adjacent boreholes to ensure a logical and geologically feasible profile. As indicated at the beginning of this section, maintenance of cross sections (even simple or diagrammatic) is recommended. Consideration of the implication of any anomalies in the context of the ground model requires checking to ensure that the sequence being recorded broadly fits what was anticipated. Any anomalies that are identified through this process should be addressed. This may mean that revisions are required to the ground model.

16.8 Checking against test results

Once a field record has been compiled there is a perennial question about review of the descriptions in the light of further information, most commonly in the form of laboratory test results. The simple and common sense rule is that the log must always be reviewed in the light of other information and test results. The initial purpose of this is to ensure that the logger's thumb or hammer swing is appropriately calibrated in terms of the preparation of the descriptions. It is hoped that any miscalibration would have been picked up at an earlier stage in the process and certainly not later than the check logging (see Section 16.4).

However, it is very important only to change these primary records when the logger is found to be in error. The scope for making changes otherwise is quite limited as indicated below:

- Checking the particle size grading of coarse primary or secondary fractions is often necessary to ensure that the logger's eye is calibrated to make the volume to mass conversion when describing particle size distributions. If a particle size grading shows that the reported sand and gravel proportions are in error (this is a common difficulty) the log description could be revised. However, it is important to check the test result against the specific sample description not just the overall stratum description. In addition, account needs to be taken of any soil fabric present, such as clay laminae or sand pockets. In the absence of any such mitigation, the description should normally be adjusted to accord with the laboratory test result.

- Classification tests (liquid limit and plasticity index) can in theory be used to check the distinction in fine soils between clay and silt. The A line on the plasticity chart is not very reliable in distinguishing between fine soil types but does give a broad indication of the plasticity of the soil in this regard. As above, the cross check should be made with the sample description rather than the stratum description. The field log should only be changed if the reviewer is assured that the logger is incorrectly calibrated. If this is necessary, the reviewer might like to consider giving the logger a refresher on the hand tests used to make this distinction in the field.

- The occurrence of apparent discrepancies between the assessment of consistency in fine soil or the manual strength assessed in rock and the laboratory measured strengths is not uncommon. This does not mean that the description is wrong. First, it should be remembered that the manual tests carried out in the field by the logger assess the strength of the fissure or joint bounded blocks and report the material strength. Once other tests are carried out on larger specimens, the discontinuities affect the result which is thus some measure of the mass strength depending on the sample size and the fracture spacing. Second, the manual tests carried out in the field to determine consistency as part of the description, together with the field tests (such as vane, point load) and laboratory tests (uniaxial or triaxial) all subject soil or rock specimens of different size to different failure modes at different strain rates and at different scales. It would therefore be surprising if the results did agree. It is thus a matter of reporting the assessments as made. Discrepancies are not necessarily a matter of changing the field log to smooth over the discrepancies; rather this is a matter that needs to be discussed in the geotechnical investigation report (GIR) as required under EN 1997-2.

16.9 Editing the field log to completion

A good field log conveys to the reader a clear and unambiguous image of the soils and rocks seen. This requires a balance between conciseness and comprehensiveness in the main description. The use of main and detail descriptions allows presentation of the overall description together with depth-related information on local deviations.

Before completion of the borehole log, the descriptions should be proof read against the original information, checking that the log says what it was meant to say and that it describes the ground before the drilling was carried out. The logger should also check for correct spelling and grammar.

Any reader of a borehole log will carry out these checks, explicitly or even without trying, in reviewing or just using the logs. If errors are found as a result, the logger has fallen down on their responsibilities. The logger should be carrying out these checks mentally during every stage of the description process, whilst logging

TIP ON INTERNAL CONSISTENCY CHECKS

The logger should carry out checks for internal consistency on the completed log. These checks could include:

- Do the consistency/density descriptors reflect the blow counts?
- Are the water strikes consistent with the strata boundaries and descriptions, so that the water entry is from the sand not the clay?
- Is the gravel angular because it is cobbles broken by chiselling?
- Is the clay firm in situ or has it been softened by chiselling and/or water addition?
- Do the density and particle size grading descriptors reflect the chiselling times and levels?
- Is the 'pushing boulder' comment by the driller included as a remark and/or detail description and at the same depths as the chiselling?
- Have the descriptions been checked against the test results for correctness, or at least robustness if apparently inconsistent?
- Does the gravel come from hole collapse during drilling in of casing?
- Is the mudstone weak in situ or has it been softened by flush water?
- By definition, the following statements must be true:
  - TCR ≤ 100%
  - SCR ≤ TCR
  - RQD ≤ SCR
  - SCR < TCR then If minimum = NI
  - RQD = 100% then minimum If ≥ 100 mm
  - RQD = 0% then maximum If ≤ 100 mm
- Description of spacing of joint sets should agree with If
- Average If (and range) is broadly related to RQD (Priest and Hudson, 1976).

the samples, compiling the stratum descriptions, writing up the field log and double checking whilst proof reading the log.

Checks for internal consistency are most important to maintain the credibility of the logger. Readers who will be finding these simple errors will include your project supervisor, your client and carry through to the lawyer facing you in court if you get it wrong.

Over the years, a number of editorial problems have been solved in the preparation of field logs where the original description was not clear or was ambiguous. Examples of the resolution of this difficult wording is given in Table 16.3. These examples pick up a number of the issues highlighted earlier in this section.

The field log is now nearly finished. The typed or printed log should be reviewed by the logger with great care. The skeleton log and descriptions may have been input by someone unfamiliar with the site or the nuances of the driller or the logger.

The logger is the person responsible for the final product so it is up to them to carry out a final cross check with the original field records for correctness and internal consistency. The log should be checked to ensure that all field observations including driller's observations, logger observations on samples and from borings, trial pit and face records and all related observations such as hole stability, ease of excavation, groundwater inflows and material variability have been included.

At this point the field borehole or trial pit logs have been compiled with the primary record at this stage being the printed log. This used to be the fundamental output from any ground investigation. However, current procedures require

Table 16.3 Editorial hints in compiling descriptions

Range observed	Terminology
N values vary from 9 to 35 in a random pattern	Generally medium dense
N values vary from 2 to 20	Medium dense SAND. Probably affected by piping at 8.5 m
Consistency increases gradually	Soft becoming firm below 5 m
Colour varies randomly from dark grey to greyish brown to light brown	… grey and brown …
A sand is fine, medium and coarse, but two samples are slightly gravelly	SAND
Gravel size fragments of weak sandstone with sand	Very dense probably weakly cemented SAND
Uncertain of clay consistency in situ because of sample disturbance	Possibly soft CLAY
Sample is small bag of gravel, chiselling recorded, standard penetration test refusals	Probably yellow weathered SANDSTONE Recovered as strong gravel size fragments
Undisturbed samples not examined under client instructions	'Description based on disturbed samples only' in Remarks

that the data is all entered into a database using relevant computer programs. It is intended that this data should form part of the project database which will also include all the other information such as the results of field and laboratory tests.

This emphasis on how the data is managed is changing the investigation process so that the written field log is only one part of the investigation output. It still forms and will continue to form, for the foreseeable future, the fundamental product that the site investigation company sells.

It is always possible to print out a paper field log from this database, but the emphasis now is that it is the data that are important in preference to the actual field log itself. While this is true, all the comments above on the process of preparing a sample description, then a stratum description and then a field log and the checking of that record for completeness and accuracy remain the same. It remains the responsibility of the logger to check that all the data used to make up the field log is correct, complete and internally consistent.

An increasing number of projects require the database to be passed to the client (or consulting engineer) for use in their analysis and design. The method of transmitting this data is to produce the data in some ASCII format such as the AGS format (Association of Geotechnical and Geoenvironmental Specialists) in the UK or the new DIGGS data format (Data Interchange for Geotechnical and GeoEnvironmental Specialists) internationally. Whichever digital data format is used, it includes transmission of the geological descriptions, together with all other information on the field log and all other investigation observations and test results.

17 Description Process – Field Exposures

The logging of boreholes has certain advantages and limitations compared with logging field exposures. Some of the pros and cons of each approach are summarised in Table 17.1. The term 'exposure' throughout this section is taken to include the whole range of in situ observation opportunities, listed in Table 17.2.

Complete logging of the soil or rock fabric including, in particular, bedding and discontinuities is best carried out in the field using in situ exposures. Within reason, the larger the exposure that can be used the better, although obviously practical or safety limitations which mean that most exposures used in investigations are at most a few metres in dimension. This is still a huge improvement on borehole samples whose diameter is usually less than about 100 mm.

The main advantage offered by exposure logging is that the advantages outlined in Table 17.1 become available for a comparatively low cost. Trial pits are a very

Table 17.1 Comparison of borehole and exposure logging

Criterion	Boreholes	Exposures
Depth	• Unlimited in theory, but certainly sufficient for all investigation needs.	• Limited to less than 10 m unless shafts or adits are used (expensive).
Access to in situ material	• Poor, all materials are examined indirectly after recovery. • One dimensional scan line sample available.	• Excellent once effects of excavation removed. • Two or three dimensional sample available. • Amount of material exposed can be constrained by health and safety measures.
Disturbance	• All samples presented have been disturbed to some extent. • Amount of disturbance can be very significant and preclude assessment of in situ condition.	• Little or none
Samples	• High quality samples possible if appropriate methods in use. • Dimension of samples is generally <0.1 m. • Volume of samples very limited.	• Highest quality samples possible. • Dimension of samples greater than 1 m, or can be much larger. • Large volume of sample possible.
Groundwater	• Progress not limited by water inflows. • Continuation below water table allows investigation of changes with depth.	• Generally curtailed by water inflow, even if pumping is allowed for. • Not usually practicable to continue further to investigate groundwater profiles.

economic method of investigating the material in the top 3–5 m below ground level by providing good exposure and in situ testing and sampling opportunities. Despite, or because of, these advantages and disadvantages, each of these methods remains a method in the armoury.

17.1 Geological and geomorphological mapping

The mapping of geological conditions on site is particularly important in order to confirm the geological units present as indicated on the geological map as well as to identify the situation of the site in the sense of natural and anthropogenic processes that have affected the ground and could affect the proposed works. Recording the conditions on the site at an early stage will often be critical in determining the success of all stages of investigation. A large majority of site failures are marked by a failure to carry out all stages of the site appraisal thoroughly such as desk study, appreciation of the geomorphology, mapping and appropriate forms of investigation (Hutchinson, 2001). The requirement for the desk study to include geological, engineering geological and hydrogeological mapping is explicitly called for in EN 1997-2.

The procedure for carrying out geological mapping is not covered in this book as this element of geological recording is still taught in undergraduate courses. Most geology graduates are capable of producing a map showing the geological units and their arrangement. This will still require checking in the same way as borehole logging, but less prior training by the investigation company should be necessary. Guidance on carrying out of geological mapping for engineering purposes is given in Anon (1972) and IAEG (1981).

Mapping the ground for engineering purposes usually includes the need to recognise and record geological processes. This used to be referred to as mapping the superficial deposits, but is now generally termed geomorphological mapping with an emphasis on recording features showing process as well as arrangement. The guidance papers and reports that indicate how this should be done have been compiled into a single volume by Griffiths (2002). Detailed guidance on reading the ground is also given in the fourth Glossop Lecture by Hutchinson (2001). Not all geology graduates will have been taught this aspect of mapping and guidance may be required to get them started in practice.

17.2 Field logging

Field or geological logging enables strata in an exposure in situ to be recorded with minimal disturbance and in three dimensions (3D).

There are two aspects of the ground that are available for recording in exposures, as outlined in Table 17.2. These are the disposition of the geological units in terms of the lithologies present and any geological structure that affects their arrangement. The second is the condition of these units in terms of their material and mass properties; it is the potential for complete recording of the mass properties that benefits most from the generally larger scale available in an exposure.

The logging of the geological soil or rock units present has, in effect, been covered in Chapters 3 to 14, as there is no difference in principle between logging an exposure and a borehole. There will be more and, we hope, better quality information available in a field exposure to allow fuller recording of the ground

Table 17.2 Types of exposure

Position of ground or exposure	Logging method	Examples
Horizontal surface	• Geological mapping • Geomorphological mapping • Fracture logging • Scan lines for fracture logging • Fracture logging	• Ground surface • Areas of site preparation • Foundation excavations
Vertical surfaces	• Geological logging • Geomorphological mapping • Scan lines for fracture logging	• Cliffs • Quarry faces • Excavations • Cut slopes
Surface excavations	• Geological mapping or logging • Fracture logging • Scan lines for fracture logging	• Trial pits • Trial trenches • Trial shafts
Underground excavations	• Geological logging • Fracture logging • Scan lines for fracture logging	• Adits • Shaft floors • Shaft walls • Tunnel faces • Tunnel walls
Investigation of existing foundations	• Disposition and condition of structural elements in three dimensions • Geological logging • Fracture logging	• Trial pits • Trial trenches

conditions, although this is not always the case. The recording process is carried out in accordance with the principles laid out in Chapter 16, although working under cover with benches is unlikely to be appropriate.

The initial problem that will face any logger setting out to record an exposure in the field is the sheer quantity of information that is available. This requires help to be provided in two ways. The first is to guide the logger in breaking the task down into manageable elements in the field and will include the steps described below.

Before going to site to record anything, the project leader should inform the logger of the level of detail that is to be recorded and the scale at which any face logs are to be prepared in the field. For instance, is this a 2, 6 or 10 trial pit per day operation, or is this a failure investigation where all information should be recorded in case it might be critical? Note that in exposure logging, as in logging activities, the level of detail and the level of quality are different things. A simple log is not detailed but can be of high quality because all the information that really matters is presented accurately, therefore:

- Encourage the practice of examining the arisings at the bucket or spoil heap during the pit excavation in order to start compiling a mental image of the strata and the outline descriptions at an early stage.
- The first exercise in the field should be to record the disposition of the units, be these soil layers, fractures or other features.

- The second exercise is to describe the units or features observed.
- Encourage the logger to describe a single face of the trial pit first and then to look at the other faces in order to record variations.
- If the face is large, break it up into squares and then record the contents of each square in turn. The size of the squares will vary from 1 to 2 m in a detailed fabric logging exercise, to 5–10 m in an excavation face to 20–50 m in the case of a length of high cliff. Division into squares can use strings (colour coded if necessary) or paint marks on the face.
- Assist the accuracy of the sketch by surveying a number of marker points, or the strings or paint marks.
- Alternatively, create a photo-montage of the face before carrying out the description to act as a base map, as described below.

The second means of helping a logger is to provide them with appropriate field recording proformae which will enable them to begin sketching the features or filling in the boxes provided with data which it has been decided in advance are necessary. Some example proformae are provided in the Appendix, but it is probable that individual companies and projects will evolve their own forms for use in the particular projects that they carry out. It is important to check the information included on these forms with the schedule of information that the standards and specifications call for. This may require the addition of extra boxes or fields to ensure that the logger records all the required information in the field.

Having recorded the soils and rocks present in the exposure, the next and usually most important step is to record the mass characteristics and particularly the discontinuities. The discontinuity recording taught to geology students may have included taking readings on a number of discontinuities in order to establish the discontinuity sets. This will often have been in the context of structural mapping rather than for engineering purposes. These readings will have been taken wherever exposure or access permits, but not in any systematic manner. The aim of the systematic approach used in engineering investigations is to obtain a sufficient quantity of readings in order to be able to plot a stereograph and to analyse the data for trends or concentrations.

In the detailed stages of an investigation, including those carried out as part of construction monitoring, it is usual and preferable to carry out the recording of discontinuity orientation and other characteristics in a systematic manner through the setting out of a number of scan lines. There should preferably be a minimum of three mutually perpendicular scan lines, of a length suited to the size of the exposure. It is generally suggested that scan lines longer than about 10 m or 20 m are more easily handled if they are subdivided. It should be noted that the scan lines should all be of about the same length. There is little point carrying out long scans in the horizontal directions because it is easy, but with very limited vertical scans because access is difficult. Guidance on this approach for suggested procedures is provided in detail by ISRM (1981).

Having set out the scan lines, a record is made of every fracture that crosses the line. The information to be recorded is in full accordance with the guidance set out in Chapter 11. An example proforma record sheet for use in recording discontinuities that intersect a scan line is given in the Appendix. This sheet can also be used for the systematic recording of fractures in a borehole core in those cases where a record of every fracture in the core is called for. In these circumstances, not all boxes can be filled in, but the principle remains valid.

A majority of engineering investigations lack the sort of exposure to make this full three-dimensional approach possible. The normal exposure available might be a large trial pit where sufficient rock or soil is exposed to allow a reasonable number of observations of discontinuities to be made, but not through any rigorous scan line method. Even more common is the single scan line approach afforded by a rotary core.

Where orthogonal scan lines are not available, the set of observations will not be properly representative of the in situ conditions. For example, most rotary cores are taken in a vertical direction and will completely sample the horizontal fracture set, most commonly the bedding, but will be totally inadequate in sampling the vertical fracture sets. In such cases the use of inclined cored holes should be considered. However, these will be of little benefit unless some sort of orientation of the hole or core is carried out. This will either comprise orientation of the core before drilling, or some form of down hole inspection by camera, televiewer or impression packer.

17.3 Information to be recorded in exposure logging

The information that should be recorded in logging field exposures is listed in Table 17.3 with the requirement being cumulative from row to row of the table.

It is strongly recommended that all field logging is carried out on preprinted proformae. Forms for recording field logging information can be general forms, such as for trial pits, or forms developed for a specific project given the amount and level of detail of information to be recorded. This is because the proforma should have been designed to include boxes that all need to be filled in so as to provide a complete record (as detailed in Table 17.3). Not least this will include dates, personnel and a record of which Standard procedure was being followed (a mandatory requirement according to EN 1997-2). This information is easily forgotten if a notebook or blank sheets of paper are used. In addition, most field loggers are not good at placing their field notebook into the project file at the office. The proforma is readily signed off and filed so this is very useful during log checking and if there is ever any question about the logging.

There are two additional activities which are essential to all exposure recording that should be included in project plans from the outset. These are survey control and safety.

TIP ON LOGGING IN WET CONDITIONS
- When logging or weather conditions are very wet an alternative approach is to use waterproof paper or laminated proformae. The base proforma for the project can be laminated to make it waterproof. The logging in the field is carried out using waterproof tools such as underwater pens or Chinagraph pencils.
- On returning to the site office, this record can be photocopied for the file. The laminated sheet is then cleaned off and re-used.
- This approach is particularly useful in mapping underground works which can be very wet.

Table 17.3 Summary of information to be recorded on exposures

Location/type of exposure	Information required	Procedure/actions
General	• Use logging proforma because it contains the prompts that allow all relevant information to be recorded.	
Trial pits, trenches	• Dates of excavation and logging • Description of each stratum • Depth of stratum changes • Log as many faces as needed to record variability • Sketch geology in each face unless they are all the same • Face stability • Description and location of all discontinuities • Levels of water inflow, estimate of rate of flow, rest water levels if achieved • Details of any pumping carried out • Equipment in use including excavator • Subjective ease of excavation • Weather • Plan dimensions and orientation • Location of pit or survey marker • Orientation of long axis of pit or trench • Whether logged in situ or on arisings in bucket • Identify continuity around all four faces • Depth and position of all samples taken • Depth and position of all tests carried out	• Faces are logged by examination of the exposed faces, usually from the surface • Materials are logged on arisings in the bucket • Logging in situ requires the excavation to be supported before the faces can be cleaned of smeared and disturbed materials • Samples are taken from selected strata (ensure samples are large enough for proposed testing) • Tests are carried out at required depths or in selected materials, either in situ or on arisings in the bucket • Log faces or arisings, not the recovered samples • Photograph pit in two directions • Include visible scale in all photographs (hammer, ranging rod or survey staff) • Photograph spoil heap(s)
Foundation inspections	• Measure and record size, depth and arrangement of foundations • Record location in plan and section related to a grid or permanent markers • Type of foundation construction, e.g. brick, concrete, wood • The presence and nature of any blinding layers or sealing materials • The underlying and surrounding materials should be described	• A plan view is essential • The location should be referenced to permanent features or site grid • Sections should be drawn to show the various elements clearly • Dimensions should be included on plans and sections to provide legibility and accuracy • Photograph the features revealed by the excavation • Photograph the location of the excavation

Table 17.3 *Continued*

Location/type of exposure	Information required	Procedure/actions
	• Describe structural features such as expansion joints, weep holes, ties, damp proof course • Describe any defects or areas of poor quality construction • Describe cracks and joints individually showing aperture and variation in crack width along the length of the crack, infill, surface alteration/staining • Note changes in brickwork colour which may indicate a change from engineering bricks to standard type bricks	
Large excavated or natural exposures	• Record disposition of different materials • Record the material and mass characteristics of all strata • Record any evidence of natural processes affecting the ground or the site	• Create some form of grid to break face into regular units that can be mapped (see below)

17.4 Surveying

Exposures that are logged as part of an investigation may be long or short lived. Nevertheless, the need to record their location in three dimensions is fundamental. This is no different from the need to survey the position of boreholes as part of the investigation record. Ideally this surveyed position would be to the national grid system, although a site grid system is also acceptable. In addition, recording the position relative to some other (permanent) landmarks or structure is also recommended.

The inclusion of survey control becomes more of a problem on large faces such as quarry or cliff faces or trenches. In these cases the survey team will be able to provide a number of reference points around which the logger can build the face log. Comments are given above on dividing the exposure into squares using string or paint markers and additional sophistication can be achieved by labelling the string lines with reference details such as chainage or reduced levels. If these labels can be made legible in the photographic record of the face, it makes life easy for everyone.

In recent years various remote surveying methods have become more readily available at a reducing cost. These methods are generally suitable for recording geological exposures and are thus particularly useful where it is becoming increasingly difficult to make a safety case for spending several days at the foot of, or on, the cliff without special methods of access and protection.

These methods include the use of aerial photography, surveying by reflectorless total station, digital photographic face mapping and the use of a laser such as the system known as Light Detection and Ranging (LiDAR).

Aerial photography is a well established method of mapping horizontal surfaces, although clearly less suitable for vertical or short term exposures. For large sites this method can be highly cost effective with appropriate images available for only a few hundred pounds/euros/dollars.

The three-dimensional surveying of faces using a reflectorless total station is a suitable method of creating a terrain model of a face. This is a method particularly useful in quarry reserves estimating, where the model can be evolved by the inclusion of geological information into volume calculations.

The recording of faces using conventional stereo photographic and photogrammetric methods is well proven. However, they are manual and time consuming and are becoming increasingly rarely used. The reason for the decline in the use of these methods is that there is still a need to go the face and record a grid of survey points to allow the appropriate rectification to be carried out.

Terrestrial digital photogrammetry allows acquisition of high resolution digital photographs of a target face in stereographic pairs and the input of limited survey control allows the construction of digital face drawings. From these drawings it is possible to calculate the orientation of flat planar surfaces and linear traces. The equipment used can be sophisticated and expensive, or off the shelf and inexpensive using modern cameras. The software is also available for purchase. This sort of approach is particularly good news for three reasons:

- First, it allows systematic recording of fractures over the whole face, even when that face is high and would therefore often be inaccessible.
- Second, it allows recording of the orientation of a large number of discontinuities, accurately and economically.
- Third it avoids the safety concerns about working at the foot of a rock slope.

An example of the type of output from such as approach is shown in Figure 17.1. A review of the current methods of terrestrial photogrammetry is provided by Haneberg (2008).

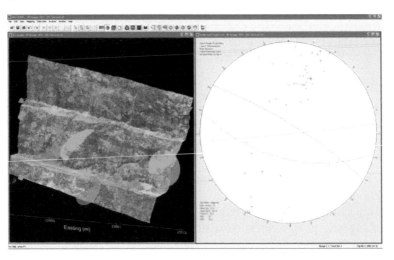

Figure 17.1 Image of a face with the interpreted discontinuity locations and orientations ready for analysis of slope stability (image courtesy of Commonwealth Scientific and Industrial Research Organisation, CSIRO).

Table 17.4 Safety considerations in field logging

Safety considerations in trial excavations include:

Working methods
- Stand at the short end of the trial pit as it is most stable.
- Stand well away from the excavator.
- Stand so that driver can see you.
- Do not move from your position until the driver has acknowledged your intention and the bucket is on the ground. You may then approach the pit for inspection or measurement.
- The excavator driver is your banksman – do not release him to dig next pit.

Entry to trial excavation
- Do not enter any excavation until a risk assessment has been carried out and all identified measures are in place.
- An excavation in this sense is one where your nipples would be below ground level. So if you remain standing up, the allowable depth is greater. This is an unlikely scenario and the risk assessment should assume you are on all fours logging or sampling.
- Notwithstanding the risk assessment, it is up to the logger to look at the excavation and confirm the safety before entering the pit.
- IF IN DOUBT DO NOT ENTER PIT.
- Mitigation measures are likely to include:
  - support
  - access and escape routes (including a ladder)
  - monitoring for gas
  - banksman on surface.

A variety of laser scanning tools are available for face mapping. The platforms for these can be as varied as the imagination including airborne and ground borne vehicles. Laser scanning provides, at low cost, a very high density of data and rapid data collection. As with the other methods a 3D digital terrain model (DTM) is created. This has been used in a variety of applications. In the context of exposure logging, a map is created of the exposure which the logging geologist can use as a base for the addition of records and observations with no or at least only limited close access to the face being required.

17.5 Safety

The safety of all personnel involved in investigation is, of course, a basic and essential requirement. Field exposure logging is a potentially dangerous activity and the risks should be assessed and steps taken to mitigate those risks before any field work is carried out. The requirement for awareness of safety falls into the categories of:

- working in or near excavations that are existing or have been dug
- working below high faces
- working with investigation-related plant
- working near plant associated with the construction or quarry works
- lone working.

The potential for risks associated with trial pitting in terms of plant, pit collapse and gas are supposedly well understood. However, injuries and fatalities continue to occur as a direct result of trial pitting. It is therefore clear that more remains to be done to protect our loggers in these circumstances. Given that the logging of and possible entry to trial pits is one of the most common activities and potentially the most significant hazard loggers will be subjected to, some of the safety considerations for trial pits or trenches are summarised in Table 17.4.

The risks associated with other activities, such as face logging, are probably less well understood. The reason that injuries are less apparent is that such logging is much less frequently carried out within site investigations. It is likely that all geologists have happily inspected excellent cliff exposures without carrying out a risk assessment, even if only in their undergraduate mapping days. What use is a hard hat if a falling particle is a small cobble that has fallen more than 10 m? This is a plea for appropriate risk assessments to be carried out before any field logging activity.

In the modern world of safety control through a system of risk assessments, alternative methods of access need to be provided. If one of the remote survey or recording methods discussed above is not possible, then access to the face will be obtained through some form of roped access or by using access plant such as mobile platforms.

Appendix: Pro-Forma Field Record Sheets

In this Appendix, a range of field record sheets for use when describing soils and rocks is provided. The intention of these sheets and their layout is to act as a prompt for the information that should be recorded in the execution of such activities.

The information needed to complete the fields on the sheets includes the information that the Standards call for in the reporting of such exposure recording. It is not the case that the layout of the sheets is standardised, far from it. However, the information that is to be recorded is common and these examples are provided for just that purpose.

The sheets that are provided on the following pages are:

- Sample description for the systematic description of samples such as from a borehole. This sheet can also be used to record the description of any other *ad hoc* sampling exercise.
- Sample description for the systematic description of borehole samples which includes space to record the stratum boundaries and the descriptions of the strata. This format is very useful for field use but does provide less space for the sample descriptions than the previous sheet.
- Extruded sample description for use when carrying out extrusion and splitting of samples for detailed recording of fabric features. This is most commonly carried out in detailed investigations of recent normally consolidated soils, but the format can be used on any core sample of up to 1 m in length.
- Core description to record the description of continuous rotary cores of soil or rock including the descriptions, discontinuities and fracture indices.
- Borehole discontinuity log for use when recording detail on each discontinuity within a core and when the requirement is to record all available information on each fracture in turn. This amount of information does not readily fit onto a core description sheet, or onto a normal field log and so this information is tabulated and entered into the field records as a table.
- Trial pit log for recording the soil or rock exposed in trial pits or short trenches. Long trenches will require a rather larger sheet of paper. The boxes available on this form include those for recording samples and field tests carried out in situ. This form can also be used for the mapping horizontal surfaces such as foundation excavations or trench bases exposed during the course of the construction works.
- Exposure log for use in *ad hoc* exposures or excavations such as may typically be found in exposures during construction works.
- Scan line fracture log for use in the systematic recording of fractures on scan lines set up in the field. This sheet is similar to the borehole discontinuity log but has additional columns owing to the extra information that is available for recording in a field exposure.
- Tunnel face log is an adaptation of several of the other sheets and is suitable for use in a machine or hand bored circular section tunnel. The shape of the face is readily adjusted for other profiles. This is the sort of sheet that can often sensibly be laminated for use underground.

These sheets are presented here for information and use as readers wish. The forms are available on line for free download as pdf documents on the author's web site at www.drnorbury.co.uk.

D	David Norbury Engineering Geologist	N				Sample Description

Project				Project No		Borehole No	
Date				Described by		Sheet No	

Sample No	Type	Depth m		Recovery mm	Description of samples
		From	To		

Sample Types
D   Small disturbed sample   TkW   Thick walled open sample
B   Large disturbed sample   TW   Thin walled open sample
Blk Block sample   PS   Piston sample
Note quality of recovery if   CS   Core sample
significantly disturbed   Note sample diameter if not 100 mm

Remarks

Standard in use: 5930 Amdt1; 5930 pre Amdt 1; Other

Sample Description with Stratum description

D David Norbury
Engineering Geologist **N**

Project				Project No		Borehole No	
Date				Described by		Sheet No	

Sample No	Type	Depth, m		Recovery, mm	Description of samples	Stratum description (include depths of boundaries)
		From	To			

Sample Types
D   Small disturbed sample
B   Large disturbed sample
Blk Block sample

TKW Thick walled open sample
TW   Thin walled open sample
PS   Piston sample
CS   Core sample

Remarks

Standard in use: 5930 Amdt1; 5930 pre Amdt 1; Other

| D | David Norbury Engineering Geologist | N | **Extruded Undisturbed Sample Description** |

Project	Project No
Described by	Borehole No
Date	Sample No	Depth of top

Description	Sketch of bedding, fabric and discontinuities	Depth from top of sample

Sample Type:	Remarks
TkW Thick walled open sample	
TW Thin walled open sample	
PS Piston sample	
CS Core sample	Standard in use: 5930 Amdt1; 5930 pre Amdt 1; Other

D David Norbury Engineering Geologist N			Core Description							

Project			Project No							
Described by			Borehole No							
Date			Sheet No							

Depth m	Description		Core Run		TCR m	SCR m	RQD m	If		Min Mode Max
	Main	Detail	From	To				From	To	

Samples taken / tests performed	Remarks
	Standard in use: 5930 Amdt1; 5930 pre Amdt 1; Other

Borehole Discontinuity Log

D **N** David Norbury
Engineering Geologist

Project No
Project Name
Hole ID

Date
Logged By
Checked By

Discontinuity reference	Top m	Base m	Set Number	Type	Dip deg	Dip Dim deg	Small Scale Roughness (Rough, smooth, striated)	Intermediate Scale Planarity (Stepped, Undulating, Planar)	JRC	Wall Strength MPa	Surface appearance (eg polished)	Aperture Observation (eg tight, closed, open, infilled)	Infilling Material (eg clay)	Wall Weathering
1	1.00	2.00	3	BF	45	080	RO	PL	8	10	P	infilled	clay	II

Standard in use: 5930 Amdt1; 5930 pre Amdt 1; Other

D	David Norbury Engineering Geologist	N	**Trial Pit Log**

Project	Project No	Trial Pit No
Date	Described by	Sheet No

Face Sketch (Indicate stratum boundaries and face(s) sketched, use additional sheets as required)

Samples and Tests			Strata	
Depth m	Type & No	Test results	Depth m	Description

Groundwater Observations (level and rate of inflow)		
	Strike of Face A	deg
	Length of Face A	m
	Width of Face B	m
	Weather	

Plant in use	Record of photographs taken
Stability	
Support in use	

Sample and test types		TkW	Thick walled open sample	Remarks
D	Small disturbed sample	TW	Thin walled open sample	
B	Large disturbed sample	PS	Piston sample	
Blk	Block sample	CS	Core sample	
V	Vane test		Note sample diameter if not 100 mm	

Standard in use: 5930 Amdt1; 5930 pre Amdt 1; Other

Exposure Log

D David Norbury
Engineering Geologist **N**

Project Name		Project No	
Exposure Reference		Logged By	
Exposure Location		Date	
		Sheet	
Exposure Type & Details	ie Natural/Man Made (eg cutting), Slope angle, Orientation, Excavation method, Support, Height, Other comments		
	Standard in use: 5930 Amdt1; 5930 pre Amdt 1; Other		

Chainage m	ROCK MASS DESCRIPTION	SKETCH OF EXPOSURE	REMARKS incl. photos & samples

Scanline Log

D David Norbury
Engineering Geologist **N**

Project No
Project Name
Scanline Location

Date
Logged By
Checked By

Discontinuity reference	Start m	End m	Set No	Type	Dip deg	Dip Dirn deg	Termination (Bottom/Top) X, R or D	Small Scale (Rough, smooth, striated)	Intermediate Scale (Stepped, Undulating, Planar)	Large scale Waviness, (Curved, Straight)	JRC	Wall Strength MPa	Surface appearance (eg polished)	Aperture Observation (eg tight, closed, open, infilled)	Infilling Material (eg clay)	Wall Weathering	Seepage or flow rate (l/min)
							RD	RO	PL	CU	8	10	P				
1	1.00	2.00	3	J	45	080								infilled	clay	II	

Standard in use: 5930 Amdt1; 5930 pre Amdt 1; Other

D	David Norbury Engineering Geologist	N	Tunnel face log

Project	Project No	Face chainage/ Ring no
Date	Described by	Sheet No

Face and wall Sketches

Samples			Strata	
Depth m	Type & No	Test results	Depth m	Description

Groundwater Observations (location and rate of inflow)	Record of photographs taken
Comments on face stability	Remarks

Standard in use: 5930 Amdt1; 5930 pre Amdt 1; Other

References

Anon, 1970. The logging of rock cores for engineering purposes. Engineering Group of Geological Society Working Party report. *Quarterly Journal of Engineering Geology,* 3, 1–24.

Anon, 1972. The preparation of maps and plans in terms of engineering geology. Engineering Group of Geological Society Working Party report. *Quarterly Journal of Engineering Geology,* 5, 295–382.

Anon, 1977. The description of rock masses for engineering purposes. Engineering Group of Geological Society Working Party report. *Quarterly Journal of Engineering Geology,* 10, 355–388.

Anon, 1988. Guide to soil and rock descriptions. *Geoguide 3,* Geotechnical Control Office, Hong Kong.

Anon, 1990. Tropical Residual Soils. Engineering Group of Geological Society Working Party report. *Quarterly Journal of Engineering Geology,* 23, 1–101.

Anon, 1995. The description and classification of weathered rocks for engineering purposes. Engineering Group of Geological Society Working Party report. *Quarterly Journal of Engineering Geology,* 28, 207–242.

Anon, 2006. *Peat Landslide Hazard and Assessments.* Best Practice Guide for proposed electricity generation developments. Scottish Executive, Edinburgh.

Anon, undated. *British Standards and the Construction Industry PP 444.* BSI, London.

ASTM D2487, 2006. *Standard Practice for Classification of Soils for Engineering Purposes (Unified Soil Classification System).* Standard D2487-06e1, American Society for Testing and Materials, USA.

Baldwin, M., Gosling, R. and Brownlie, N., 2007. Soil and rock descriptions. A practical guide to the implementation of BS EN ISO 14688 and 14689. *Ground Engineering,* 40(7), 14–24.

Barton, N. and Choubey, V., 1977. The shear strength of rock joints in theory and practice. *Rock Mechanics,* 10(1–2), 1–54.

Barton, N., Lien, R. and Lunde, J., 1974. Engineering classification of rock masses for the design of tunnel support. *Rock Mechanics,* 6(4), 189–236.

Bieniawski, Z.T., 1976. Rock mass classification in rock engineering. *Proceedings Symposium on Exploration for Rock Engineering,* Johannesburg, 1, 97–106.

Bienaiwski, Z.T., 1989. *Engineering Rock Mass Classifications.* Wiley, New York.

Blackbourn, G.A., 2009. *Cores and Core Logging for Geoscientists,* 2nd edn., Whittles Publishing, Scotland, UK.

Bowden, A.J., Spink, T.W. and Mortimore, R.N., 2002. The engineering description of chalk: its strength, hardness and density. *Quarterly Journal of Engineering Geology and Hydrogeology,* 35, 355–361.

BRE, 1993. Site investigation for low-rise building: soil description. *Building Research Establishment Digest,* 383. Watford.

Brown, E.T., 1970. Strength of models of rock with intermittent joints. *Journal of Soil Mechanics, Foundation Division, ASCE,* 96, SM6, 1935–1949.

Brown, D.J., 2007. A guide to the use of volcaniclastic nomenclature in engineering investigations. *Quarterly Journal of Engineering Geology and Hydrogeology*, 40, 105–112.

BS CP 2001, 1957. *Site investigation. British Standard Code of Practice.* British Standards Institution, London.

BS 812, 1975. *Aggregate: Part 1: Sampling, shape, size and classification,* British Standards Institution, London. (superseded by 1989 version)

BS 1377, 1990. *Methods of Test for Soils for Civil Engineering Purposes,* British Standards Institution, London.

BS 1881, 1983. *Testing Concrete. Part 120. Method of determination of the compressive strength of concrete cores,* British Standards Institution, London.

BS 3882, 2007. *Specification for Topsoil,* British Standards Institution, London.

BS 5930, 1981. *Code of Practice for Site Investigations,* British Standards Institution, London.

BS 5930, 1999. *Code of Practice for Site Investigations,* incorporating Amendment 1. British Standards Institution, London.

Burnett, A.D. and Epps, R.J., 1979. The engineering geological description of carbonate suite rocks and soils. *Ground Engineering,* March, 41–48.

Cameron, A.R., 1978. Megascopic description of coal with particular reference to seams in Southern Illinois. In *Field Description of Coal,* Dutcher, R.R. (Ed.), American Society of Civil Engineers Symposium Proceedings, ASTM STP 661, 9–32.

Casagrande, A., 1947. Classification and identification of soils. *Proceedings American Society of Civil Engineers,* 73, 783–810.

Chandler, R.J., 1969. The effect of weathering on the shear strength properties of Keuper Marl. *Géotechnique,* 19, 321–334.

Chandler, R.J., 1972. Lias Clay: weathering processes and their effect on shear strength. *Géotechnique,* 22, 403–431.

Chandler, R.J. and Apted, J.P., 1988. The effect of weathering on the strength of London Clay. *Quarterly Journal of Engineering Geology,* 21, 59–68.

Chandler, R.J. and Davis, A.G., 1973. *Further Work on the Engineering Properties of Keuper Marl.* CIRIA Report No 47, London.

Child, G.H., 1986. Soil Descriptions–Quo Vadis? Geological Society, London, Engineering Geology Special Publication No 2, *Proceedings 20th Regional Meeting Engineering Group,* Guildford, 73–82.

Clark, A.R. and Walker, B.F., 1977. A proposed scheme for the classification and nomenclature for use in the engineering description of Middle Eastern sedimentary rocks. *Geotechnique,* 27, 93–99.

Clayton, C.R., 1995. *The Standard Penetration Test (SPT): Methods and Use,* CIRIA Report 143. London.

Deere, D.U., 1963. Technical description of rock cores for engineering purposes. *Rock Mechanics and Engineering Geology,* 1, 18–22.

Deere, D.U. and Deere, D.W., 1988. *The Rock Quality Designation (RQD) Index in Practice. Rock Classification Systems for Engineering Purposes.* ASTM STP 984.

Deere, D.U. and Miller, R P., 1966. *Engineering Classification and Index Properties for Intact Rock.* Technical Report No.AFNL-TR-65-116, Air Force Weapons Laboratory, New Mexico.

Department of Transport, 2007. *Specification for Highway Works.* Volume 1 of the Manual of Contract Documents for Highway Works. HMSO, London.

Early, K. and Skempton A., 1974. Investigation of the landslide at Walton's Wood, Staffordshire. *Quarterly Journal of Engineering Geology,* 7, 101–102.

Ege, J.R., 1968. Stability index for underground structures in granitic rock, in Nevada Test Site. *Memoir of Geological Society of America*, No 110, 185–198.

EN 1997-2, 2007. *Eurocode 7 – Geotechnical Design – Part 2: Ground investigation and testing*. CEN.

EN ISO 14688-1, 2002. *Geotechnical Investigation and Testing – Identification and classification of soil. Part 1: Identification and description.* British Standards Institution, London.

EN ISO 14688-2, 2004. *Geotechnical Investigation and Testing – Identification and classification of soil. Part 2: Principles for a classification.* British Standards Institution, London.

EN ISO 14689-1, 2003. *Geotechnical Investigation and Testing – Identification and classification of rock. Part 1 Identification and description.* British Standards Institution, London.

EN ISO 22475-1, 2006. *Geotechnical Investigation and Testing – Sampling methods and groundwater measurement. Part 1: Technical principles for execution.* British Standards Institution, London.

EN ISO/TS 22475-2, 2006. *Geotechnical Investigation and Testing – Sampling methods and groundwater measurement. Part 2: Qualification criteria for enterprises and personnel.* British Standards Institution, London.

EN ISO 22475-3, 2007. *Geotechnical Investigation and Testing – Sampling methods and groundwater measurements. Part 3: Conformity assessment of enterprises and personnel by third party.* British Standards Institution, London.

EN ISO 22476-2, 2005. *Geotechnical Investigation and Testing – Field testing. Part 2: Dynamic probing.* British Standards Institution, London.

EN ISO 22476-3, 2005. *Geotechnical Investigation and Testing – Field testing. Part 3: Standard penetration test.* British Standards Institution, London.

FAO, 1998. *World Reference Base for Soil Resources.* Food And Agriculture Organization of the United Nations. Rome.

Fookes, P.G. and Higginbottom, I.E., 1975. The classification and description of near-shore carbonate sediments for engineering purposes. *Géotechnique*, 25, 406–411.

Griffiths, J.S., 2002. Mapping in Engineering Geology. *Key Issues in Earth Science.* Geological Society of London.

Haneberg, W.C., 2008. Using close range terrestrial digital photogrammetry for 3D rock slope modelling and discontinuity mapping in the United States. *Bulletin of Engineering Geology and the Environment*, 67, 457–478.

Hawkins A B., 1986. Rock descriptions. Geological Society, London, Engineering Geology Special Publication No 2, *Proceedings 20th Regional Meeting Engineering Group*, Guildford, 59–72.

Hencher S., 2008. The new British and European standard guidance on rock description. A critique. *Ground Engineering*, 41(7) (July), 17–21.

Highways Agency, 2007a. *Clause 710–Testing for Constituent Materials in Recycled Aggregate and Recycled Concrete Aggregate.* HMSO, London.

Highways Agency, 2007b. *Specification for Highway Works. Volume 1 of the Manual of Contract Documents for Highway Works.* HMSO, London.

Hobbs, N. B., 1986. Mire morphology and the properties and behaviour of some British and foreign peats. *Quarterly Journal of Engineering Geology*, 19, 1–80.

Hoek, E., 1968. Brittle failure of rock. In *Rock Mechanics in Engineering Practice*, Stagg, K.G. and Zienkiewicz, O.C. (Eds.), Wiley, London, 99–124.

Hoek, E., 1983. Strength of jointed rock masses, 23rd Rankine Lecture. *Géotechnique*, 33, 187–223.

Hoek, E. and Brown, E.T., 1980. Empirical strength criterion for rock masses. *Journal of the Geotechnical Engineering Division, ASCE*, 106(GT9), 1013–1035.

Hutchinson, J., 2001. Reading the ground: morphology and geology in site appraisal. *Quarterly Journal of Engineering Geology and Hydrogeology*, 34, 7–50.

IAEG, 1981. Rock and soil description for engineering geological mapping. Report by the IAEG Commission on Engineering Geological Mapping. *Bulletin of the International Association of Engineering Geology*, 24, 235–374.

ISO 25177, 2008. *Soil quality–Field soil description,* International Organization for Standardization, Geneva.

ISRM (International Society for Rock Mechanics), 1978. Suggested methods for the quantitative description of discontinuities in rock masses. Commission on standardisation of laboratory and field tests. *International Journal of Rock Mechanics, Mining Science and Geomechanics Abstracts*, 15, 319–368.

ISRM (International Society for Rock Mechanics), 1981. Basic geotechnical description of rock masses (BGD). *International Journal of Rock Mechanics, Mining Science and Geomechanics*, 18, 85–110.

Kulander, B., Dean, S. and Ward, B., 1990. *Fractured Core Analysis: Interpretation and logging and use of natural and induced fractures in core.* AAPG Methods in Exploration Series, No 8. Tucson, USA.

Ladanyi, B. and Archambault, G., 1970. Simulation of shear behaviour of a jointed rock mass. In *Rock mechanics–Theory and Practice, Proceedings 11th Symposium on Rock Mechanics,* Berkeley, 1969, Society of Mining Engineers AIME, New York, 105–125.

Landva, O. A. and Pheeney, P.E., 1980. Peat fabric and structure. *Canadian Geotechnical Journal*, 17, 416–435.

Lauffer, H., 1958. Gebirgsklassifizierung fur den Stollenbau. *Geologie und Bauwesen*, 24, 46–51.

Long, A.J., Innes, J.B., Shennan, I. and Tooley, M.J., 1999. Coastal stratigraphy: a case study from Johns River, Washington. In *The Description and Analysis of Quaternary Stratigraphic Field Sections*, Technical Guide No 7. Quaternary Research Association, London.

Lord, J.A, Clayton, C.R.I. and Mortimore, R.N., 2002. *Engineering in Chalk.* CIRIA Report C574, London.

Macklin, S., Manning, J. and Lorenti, L., 2009. *The Engineering Geological Characterisation of the Barzaman Formation, Coastal Dubai, UAE. Part 1 – A review of the geology and classification,* in press.

Marinos, P. and Hoek, E., 2000. GSI – A geologically friendly tool for rock mass strength estimation. *Proceedings of GeoEng 2000*, Vol 1, ISRM, Melbourne.

McCook, D.K., 1996. Correlations between simple field test and relative density–test values. *Journal of Geotechnical Engineering*, 122, 860–862.

Milliman, J.D., Mueller, G. and Foerstner, U., 1974. *Marine Carbonates – recent sedimentary carbonates, Part 1.* Springer-Verlag, Berlin.

Mohamed, Z., Rafek, A.G. and Komoo, I., 2006. A geotechnical characterisation of interbedded Kenny Hill weak rock in Malaysian wet tropical environment, *Electrical Journal of Geotechnical Engineering*, 11.

Mortimore, R.N., 1990. Chalk or chalk? *Proceedings of the International Chalk Symposium,* Brighton Polytechnic, 4–7 September 1989, Thomas Telford, London, 15–46.

Norbury, D.R., Haydon, R.E.V. and Hartingdon, J.A., 1982. Rock excavation trials at Money Point Power Station. *Proceedings Conference on Earthworks*, Dublin.

Norbury, D.R., Child, G.H. and Spink, T.W., 1986. A critical review of Section 8 (BS 5930) – soil and rock description. Geological Society, London, Engineering Geology Special Publication No 2, *Proceedings 20th Regional Meeting Engineering Group*, Guildford, 331–342.

Polidoro, E., 2003. Proposal for a new plasticity chart. *Géotechnique*, 53, 397–406.

Powell, J.H., 1998. *A Guide to British Stratigraphical Nomenclature*. Special Publication 149. CIRIA, London.

Powers, M., 1953. A new roundness scale for sedimentary particles. *Journal of Sedimentary Petrology*, 23, 117–119.

Priest, S.D. and Hudson, J.A., 1976. Discontinuity spacings in rock. *International Journal of Rock Mechanics, Mining Science and Geomechanics Abstracts*, 13, 135–148.

Russell, D.J. and Parker, A., 1979. Geotechnical, mineralogical and chemical interrelationships in weathering profiles of an overconsolidated clay. *Quarterly Journal of Engineering Geology*, 12, 107–126.

Skempton, A.W., Norbury, D.R., Petley, D.J. and Spink, T.W., 1991. Solifluction shears at Carsington, Derbyshire. *Proceedings 25th Regional Meeting of Engineering Group of Geological Society*, Edinburgh.

Smart, P., 2008. Colour me sensible. *Ground Engineering*, September.

Spink, T.W., 2002. The CIRIA chalk description and classification scheme. *Quarterly Journal of Engineering Geology and Hydrogeology*, 35, 363–369.

Spink, T.W. and Norbury, D.R., 1990. The engineering geological description of chalk. *Proceedings of the International Chalk Symposium*, Brighton Polytechnic, 4–7 September 1989, Thomas Telford, London.

Spink, T.W. and Norbury, D.R., 1993. The engineering geological description of weak rocks and overconsolidated soils. *Proceedings 26th Regional Meeting of Engineering Group of Geological Society*, Leeds.

Stini, I., 1950. *Tunnelbaugeologie*, Springer-Verlag, Vienna.

Stroud, M.A., 1989. *The Standard Penetration Test – its application and interpretation. Penetration testing in the UK*, Thomas Telford.

Terzaghi, K., 1946. Rock defects and loads on tunnel supports. In *Rock Tunnelling with Steel Supports*, Proctor, R.V. and White, T. (Eds.), Commercial Shearing and Stamping Co, Youngstown, USA.

Troels-Smith, J., 1955. Characterisation of unconsolidated sediments. *Danmarks Geologiske Undersogelse, Series IV*, 38–73.

Valentine, S. and Norbury, D.R. 2010. Measurement of Total Core Recovery – dealing with core loss and gain. *Quarterly Journal of Engineering Geology and Hydrogeology*, in press.

von Post, L., 1922. Sveriges Geologiska Underskonings torvinventering ich nogra as dess hittils vunna resultat. (SGU Peat inventory and some preliminary results). *Svenska Mosskulturforeningens Tidskrift*, Jonkoping, Sweden, 36, 1–37.

Wakeling, T.R., 1970. A comparison of the results of standard site investigation against the results of a detailed geotechnical investigation in Middle Chalk at Mundford, Norfolk. In *Proceedings Conference in–situ Investigations in Soils and Rocks*, British Geotechnical Society, London.

Waltham, A.C. and Fookes, P.G., 2003. Engineering classification of karst ground conditions. *Quarterly Journal of Engineering Geology and Hydrogeology*, 36, 101–118.

Ward, W., Burland, J. and Gallois, R., 1968. Geotechnical assessment of a site at Mundford, Norfolk for a large proton accelerator. *Géotechnique*, 18, 399–431.

Index